Risk Engineering

Series editor

Dirk Proske, Vienna, Austria

The Springer Book Series *Risk Engineering* can be considered as a starting point, looking from different views at Risks in Science, Engineering and Society. The book series publishes intense and detailed discussions of the various types of risks, causalities and risk assessment procedures.

Although the book series is rooted in engineering, it goes beyond the thematic limitation, since decisions related to risks are never based on technical information alone. Therefore issues of "perceived safety and security" or "risk judgment" are compulsory when discussing technical risks, natural hazards, (environmental) health and social risks. One may argue that social risks are not related to technical risks, however it is well known that social risks are the highest risks for humans and are therefore immanent in all risk trade-offs. The book series tries to cover the discussion of all aspects of risks, hereby crossing the borders of scientific areas.

More information about this series at http://www.springer.com/series/11582

George S. Oreku · Tamara Pazynyuk

Security in Wireless Sensor Networks

George S. Oreku
Tanzania Industrial Research Development Organisation
TIRDO/North West University
Dar es Salaam
Tanzania

Tamara Pazynyuk
Far East Energy Management Company (FEEMC)
Vladivostok
Russia

ISSN 2195-433X ISSN 2195-4348 (electronic)
Risk Engineering
ISBN 978-3-319-37314-0 ISBN 978-3-319-21269-2 (eBook)
DOI 10.1007/978-3-319-21269-2

Springer Cham Heidelberg New York Dordrecht London
© Springer International Publishing Switzerland 2015
Softcover reprint of the hardcover 1st edition 2015
This work is subject to copyright. All rights are reserved by the Publisher, whether the whole or part of the material is concerned, specifically the rights of translation, reprinting, reuse of illustrations, recitation, broadcasting, reproduction on microfilms or in any other physical way, and transmission or information storage and retrieval, electronic adaptation, computer software, or by similar or dissimilar methodology now known or hereafter developed.
The use of general descriptive names, registered names, trademarks, service marks, etc. in this publication does not imply, even in the absence of a specific statement, that such names are exempt from the relevant protective laws and regulations and therefore free for general use.
The publisher, the authors and the editors are safe to assume that the advice and information in this book are believed to be true and accurate at the date of publication. Neither the publisher nor the authors or the editors give a warranty, express or implied, with respect to the material contained herein or for any errors or omissions that may have been made.

Printed on acid-free paper

Springer International Publishing AG Switzerland is part of Springer Science+Business Media (www.springer.com)

Preface

Wireless sensor network (WSN) is an area of great interest to both academia and industry. It opens the door to a large number of military, industrial, scientific, civilian, and commercial applications. They allow cost-effective sensing especially in applications where human observation or traditional sensors would be undesirable, inefficient, expensive, or dangerous. Wireless sensors have limited energy and computational capabilities, making many traditional security methodologies difficult or impossible to utilize. Also, they are often deployed in open areas, allowing physical attacks such as jamming or node capture and tampering. The threats present to a WSN and the organization of the WSN in response to these threats are influenced directly by the WSN application. As a result, WSN security design and analysis must be sensitive to this context. Otherwise, the assumptions made in the organization of the WSN and the corresponding threats may become inconsistent with the problem domain, leading to solutions that address unrealistic problems.

The security context is not a precise technical specification; rather, it is a set of security-related factors narrowing down the WSN design space to a region that is consistent with them. Clearly, conventional constraints on WSN design, such as cost, form factor, and energy must also be taken into consideration in the technical specification.

As WSN continue to grow, so does the need for effective security mechanisms. Because sensor networks may interact with sensitive data and/or operate in hostile unattended environments, it is imperative that these security concerns be addressed from the beginning of the system design. However, due to inherent resource and computing constraints, security in sensor networks poses different challenges than traditional network/computer security. They are exposed to a greater variety of attacks than other networks. The quality and complexity of these attacks are rising day by day. Information transferring through WSN needs to be protected from misuse. Modern security methods need to guarantee safety of data transmission with respect to security needs, i.e., confidentiality, integrity, and availability (CIA). Providing information security in WSN is also necessary, especially for security-sensitive applications and is one of the major concerns addressed in our proposal.

One of the main challenges is the design of these networks and their vulnerability to security attacks leading to network destruction and poor performance. Not only is the quantity and complexity of new threats increasing annually but also the appearance and the momentum. Resistance to them is becoming more and more complicated. Malicious agents are using more of these security vulnerabilities, especially to attack WSN due to the wireless security weakness.

Chapter 2. WSNs are being increasingly used in applications where Quality of Service (QoS) and low cost are the overriding considerations. With increased use, the reliability, availability, and serviceability need to be addressed from the outset. Conventional schemes of using sensor nodes and incorporating these three areas of (reliability, availability, and serviceability) to attain QoS can effectively improve not only the reliability of the overall WSNs, but also the security. We discuss the reverse look of QoS and present mathematically the three significant quality factors that should currently be taken into account in developing WSN application services and security availability, reliability, and serviceability. We also discuss specific characteristics and constraints of WSN, QoS factors when developing security applications for such networks. The security of WSNs has been addressed by providing the flow models and simulations testing using individual sensor nodes on our experiment. To evaluated possibility of establishing secure WSN through QoS, we have used Hawk nodes to demonstrate our approach experimentally. The flow models show how the QoS can be integrated to increase the security of applications running under WSNs.

Chapter 3. We develop mathematical foundations model using the barriers concept to design secure wireless sensors nodes. Security becomes one of the major concerns when there are potential attacks against sensor network nodes. Thus, we have designed fundamental security in disk-shaped to provide basic security elements that can be implemented in various sensor nodes. The mathematical models introduced are flexible and efficient so as to be embedded in sensor nodes and can create a suitable nodes components security in hostile environments.

Chapter 4. In this book, we demonstrated that the complexity of modern attacks is growing. This requires a convergent defensive strategy. Limitations in computation and battery power in sensor nodes constrain the diversity of responsive security mechanisms. We must apply only suitable mechanisms to WSN. Applications of the improved “Feistel Scheme” motivated this approach. The modified accelerated-cipher design uses data-dependent permutations and can be used for fast hardware, firmware, software, and WSN encryption systems. The approach presented shows that ciphers using this approach have less intrusion probability against differential cryptanalysis. This exceeds the currently used popular WSN ciphers such us DES and Camellia.

Chapter 5. Some special features (i.e., resource constraints, impracticality of protecting or monitoring each individual node physically, as well, as their applications normally being supported by many components such as routing and localization) of sensor networks make it particularly challenging to provide security services for sensor networks. We have described a secret distribution scheme (DSS) for sensor networks that achieve automatic secret redistribution. The goal is to support distributing the secret among new members joining a sensor network without involving a trusted agent or intervention from the user. Our analysis indicates that our new methods have some nice features compared with the previously methods. In particular, the system is efficient. Secondly, it guarantees automatic key distribution after initializations. Thirdly, it does not need urgent key distribution. Finally, it automatically interacts with nodes coalition.

Chapter 6. Current routing protocols in WSNs or even in wireless ad hoc networks are very susceptible to many attacks, i.e., stealthy attack. The most simple among these is where the adversary injects malicious routing information into the network. This results in routing inconsistencies leading to high increase in end-to-end delays or even packet losses in the network. First, we abstract two fundamental routing protocols, which can be generally grouped into two broad categories based on the intrinsic nature of WSN. We argue that none of previous proposed routing protocols satisfies all of them at the same time.

Contents

Acronyms

BS	Base station
CBR	Constant bit rate
CPB	Controlled permutations boxes
DCA	Differential crypto analysis
DDP	Data-dependent permutations
DoS	Denial of service
DSN	Distributed sensor network
DSS	Distributed signature scheme
ID	Identification
IP	Internet protocol
LAN	Local area network
LCA	Linear crypto analysis
MAC	Medium access control
MAC	Message authentication code
MTBF	Mean time between failures
MTTD	Mean time to detect
MTTF	Mean time to failure
MTTR	Mean time to repair
PC	Personal computer
QoS	Quality of service
RAS	Reliability, availability and serviceability
RFID	Radio frequency identification
SM	Security margin
TTR	Time to repair
UBR	Unspecified bit rate
WSN	Wireless sensor network

List of Figures

List of Tables

Chapter 1
Introduction and Overview

WSNs are quickly gaining popularity due to they are potentially low cost solutions to a variety of real-world challenges. They continue to grow day by day, so it needs the effective security mechanisms. Because sensor networks may communicate with sensitive data and can operate in hostile unattended environments, it is necessary that these security concerns be addressed from the beginning of the system design. However, due to inherent resource and computing constraints, security in sensor networks poses different challenges than traditional network security.

Smart environments represent the next development step in building, utilities, industrial, home, shipboard, and transportation systems automation. Like any sentient organism, the smart environment relies first and foremost on sensory data from the real world. Sensory data comes from multiple sensors of different modalities in distributed locations. The smart environment needs information about its surroundings as well as about its internal workings.

Figure 1.1 shows the complexity of WSNs, which generally consist of a data acquisition network and a data distribution network, monitored and controlled by a management center.

Their low cost provides a means to deploy large sensor arrays in a variety of conditions capable of performing a lot of both military and civilian tasks. But sensor networks also introduce severe resource constraints due to their lack of data storage and power. Both of these represent major obstacles to the implementation of traditional computer security techniques in a WSN. The unreliable communication channel and unattended operation make the security defenses even harder. Indeed, wireless sensors often have the processing characteristics of machines that are decades old (or longer), and the industrial trend is to reduce the cost of wireless sensors while maintaining similar computing power [1].

© Springer International Publishing Switzerland 2016

G.S. Oreku and T. Pazynyuk, *Security in Wireless Sensor Networks*,
Risk Engineering, DOI 10.1007/978-3-319-21269-2_1

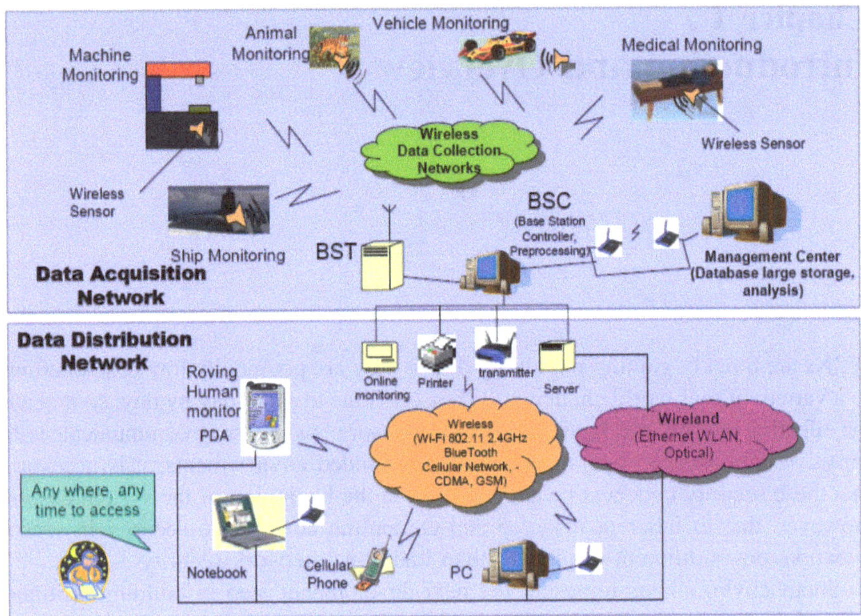

Fig. 1.1 Wireless sensor network complexity

1.1 WSN Security

Due to the resource, space, and cost constraints placed on sensor nodes in a WSN many of the security solutions for IP networks (and similar systems) are not suitable for WSN security. This, combined with a large number of threats, makes it unusually hard to build security solutions for WSN.

Security in WSN is a very active research area where new and creative solutions to the security issues are suggested on a regular basis.

The main aspects of WSN security can be classified into four major categories:

- the obstacles to sensor network security,
- the requirements of a secure WSN,
- attacks, and
- defensive measures.

1.1.1 Obstacles of WSN Security

WSN is a special network which has many constraints compared to a traditional computer network. Due to these constraints it is difficult to directly implement the existing security mechanisms to the WSNs. Therefore, to develop useful security mechanisms while borrowing the ideas from the current security techniques, it is necessary to know and understand these constraints first [2].

Very Limited Resources
All security approaches require a certain amount of resources for the implementation, including data memory, code space, and energy to power the node. However, currently these resources are very limited in a tiny wireless sensor.

- Limited Memory and Storage Space is a small amount of memory and storage space for the code. In order to build an effective security mechanism, it is necessary to limit the code size of the security algorithm.
- Power Limitation Energy is the biggest constraint to wireless sensor capabilities. Once sensor nodes are deployed in a sensor network field, they cannot be easily replaced (high operating cost) or recharged (high cost of sensors). Therefore, the battery charge must be conserved to extend the life of the individual sensor node. When implementing a cryptographic function (e.g., encryption, decryption, signing data, verifying signatures, cryptographic key storage) or protocol within a sensor node, adding security to a sensor node, the energy impact of the added security code must be considered.

Unreliable Communication
Unreliable communication is another threat to sensor security. The security of the network relies heavily on a defined protocol, which in turn depends on communication.

- Unreliable Transfer. Normally the packet-based routing of the sensor network is connectionless and thus inherently unreliable. Packets may get damaged due to channel errors or dropped at highly congested nodes. The result is lost or missing packets.

Conflicts. Even if the channel is reliable, the communication may still be unreliable. This is due to the broadcast nature of the WSN. If packets meet in the middle of transfer, conflicts will occur and the transfer itself will fail. In a crowded (high density) sensor network, this can be a major problem [3].

Latency. The multi-hop routing, network congestion and node processing can lead to greater latency in the network, thus making it difficult to achieve synchronization among sensor nodes. The synchronization issues can be critical to sensor security where the security mechanism relies on critical event reports and cryptographic key distribution [4].

Unattended Operation
Depending on the function of the particular WSN, the sensor nodes may be left unattended for long periods of time. There are three main caveats to unattended sensor nodes:

- Exposure to Physical Attacks. The sensor may be deployed in an environment open to adversaries, bad weather, and so on.

- Managed Remotely. Remote management of a sensor network makes it virtually impossible to detect physical tampering (i.e., through tamperproof seals) and physical maintenance issues (e.g., battery replacement).
- No Central Management Point. A sensor network should be a distributed network without a central management point. This will increase the vitality of the sensor network. However, if designed incorrectly, it will make the network organization difficult, inefficient, and fragile.

1.1.2 Security Requirements

As WSN is a special type of network, so it also poses unique requirements. Therefore, we can think of the requirements of a WSN as encompassing both the typical network requirements and the unique requirements suited solely to WSNs.

Data Confidentiality
Data confidentiality is the most important issue in network security. Every network with any security focus will typically address this problem first. In WSN, the confidentiality relates to the following [2, 5]:

- A sensor network should not leak sensor readings to its neighbors. Especially in a military application, the data stored in the sensor node may be highly sensitive.
- In many applications nodes communicate highly sensitive data, e.g., key distribution; therefore it is extremely important to build a secure channel in a WSN.
- Public sensor information, such as sensor identities and public keys, should also be encrypted to some extent to protect against traffic analysis attacks.

The standard approach for keeping sensitive data secret is to encrypt the data with a secret key that only intended receivers possess, thus achieving confidentiality.

Data Integrity
With the implementation of confidentiality, an adversary may be unable to steal information. However, this doesn't mean the data is safe. The adversary can change the data, so as to send the sensor network into disarray. For example, a malicious node may add some fragments or manipulate the data within a packet. This new packet can then be sent to the original receiver. Data loss or damage can even occur without the presence of a malicious node due to the harsh communication environment. Thus, data integrity ensures that any received data has not been altered in transit.

Data Freshness
Even if confidentiality and data integrity are assured, we also need to ensure the freshness of each message. Informally, data freshness suggests that the data is recent, and it ensures that no old messages have been replayed. This requirement is

especially important when there are shared-key strategies employed in the design. Typically shared keys need to be changed over time. However, it takes time for new shared keys to be propagated to the entire network. In this case, it is easy for the adversary to use a replay attack. Also, it is easy to disrupt the normal work of the sensor, if the sensor is unaware of the new key change time. To solve this problem a nonce, or another time-related counter, can be added into the packet to ensure data freshness.

Availability
Adjusting the traditional encryption algorithms to fit within WSN is not free, and will introduce some extra costs. Some approaches choose to modify the code to reuse as much code as possible. Some approaches try to make use of additional communication to achieve the same goal. What's more, some approaches force strict limitations on the data access, or propose an unsuitable scheme (such as a central point scheme) in order to simplify the algorithm. But all these approaches weaken the availability of a sensor and sensor network for the following reasons:

- Additional computation consumes additional energy. If no more energy exists, the data will no longer be available.
- Additional communication also consumes more energy. What's more, as communication increases so too does the chance of incurring a communication conflict.
- A single point failure will be introduced if using the central point scheme. This greatly threatens the availability of the network. The requirement of security not only affects the operation of the network, but also is highly important in maintaining the availability of the whole network.

Self-Organization
WSN is a typically an ad hoc network, which requires every sensor node be independent and flexible enough to be self-organizing and self-healing according to different situations. There is no fixed infrastructure available for the purpose of network management in a sensor network. This inherent feature brings a great challenge to WSN security as well. For example, the dynamics of the whole network inhibits the idea of pre-installation of a shared key between the base station and all sensors [6]. In the context of applying public-key cryptography techniques in sensor networks, an efficient mechanism for public-key distribution is necessary as well. In the same way that distributed sensor networks must self-organize to support multihop routing, they must also self-organize to conduct key management and building trust relation among sensors. If self-organization is lacking in a sensor network, the damage resulting from an attack or even the hazardous environment may be devastating.

Time Synchronization
Most sensor network applications rely on some form of time synchronization. In order to conserve power, an individual sensor's radio may be turned off for periods of time. Furthermore, sensors may wish to compute the end-to-end delay of a packet

as it travels between two pairwise sensors. A more collaborative sensor network may require group synchronization for tracking applications, etc. In [7], the authors propose a set of secure synchronization protocols for sender-receiver (pairwise), multihop sender-receiver (for use when the pair of nodes are not within single-hop range), and group synchronization.

Secure Localization
Often, the utility of a sensor network will rely on its ability to accurately and automatically locate each sensor in the network. A sensor network designed to locate faults will need accurate location information in order to pinpoint the location of a fault. Unfortunately, an attacker can easily manipulate unprotected location information by reporting false signal strengths, replaying signals, etc. In verifiable technique [8], a device's position is accurately computed from a series of known reference points; authenticated ranging and distance bounding are used to ensure accurate location of a node. Because of distance bounding, an attacking node can only increase its claimed distance from a reference point. In [9], Secure Range-Independent Localization (SeRLoc) is described. SeRLoc uses trusted locators that transmit beacon information. A sensor computes its location by listening for the beacon information sent by each locator. The beacons include the locator's location. Using all of the beacons that a sensor node detects, a node computes an approximate location based on the coordinates of the locators. Using a majority vote scheme, the sensor then computes an overlapping antenna region. The final computed location is the "center of gravity" of the overlapping antenna region [9]. All beacons transmitted by the locators are encrypted with a shared global symmetric key that is pre-loaded to the sensor prior to deployment. Each sensor also shares a unique symmetric key with each locator. This key is also pre-loaded on each sensor.

Authentication
An adversary is not just limited to modifying the data packet. It can change the whole packet stream by injecting additional packets. So the receiver needs to ensure that the data used in any decision-making process originates from the correct source. On the other hand, when constructing the sensor network, authentication is necessary for many administrative tasks (e.g. network reprogramming or controlling sensor node duty cycle). From the above, we can see that message authentication is important for many applications in WSN. Informally, data authentication allows a receiver to verify that the data really is sent by the claimed sender. In the case of two-party communication, data authentication can be achieved through a purely symmetric mechanism: the sender and the receiver share a secret key to compute the message authentication code (MAC) of all communicated data. Adrian Perrig et al. propose a key-chain distribution system for their μTESLA secure broadcast protocol [5]. The basic idea of the μTESLA system is to achieve asymmetric cryptography by delaying the disclosure of the symmetric keys. One limitation of μTESLA is that some initial information must be unicast to each sensor node before authentication of broadcast messages can begin. Liu and Ning [10, 11] propose an enhancement to the μTESLA system that uses broadcasting of

the key chain commitments rather than μTESLA's unicast technique. They present a series of schemes starting with a simple pre-determination of key chains and finally settling on a multi-level key chain technique. The multi-level key chain scheme uses pre-determination and broadcasting to achieve a scalable key distribution technique that is designed to be resistant to denial of service (DoS) attacks, including jamming.

1.1.3 Attacks

WSNs are exposed to variety of attacks as other networks. Quality and complexity of attacks are rising day by day. Attacks can be performed in a variety of ways, most notably as DoS attacks, but also through traffic analysis, privacy violation, physical attacks, and so on. DoS attacks on WSN can range from simply jamming the sensor's communication channel to more sophisticated attacks designed to violate the 802.11 MAC protocol [12] or any other layer of the WSN.

Due to the potential asymmetry in power and computational constraints, guarding against a well designed DoS attack on a WSN can be nearly impossible. A more powerful node can easily jam a sensor node and effectively prevent the sensor network from performing its intended duty.

Background
Wood and Stankovic define one kind of DoS attack as "any event that diminishes or eliminates a network's capacity to perform its expected function" [13]. Certainly, DoS attacks are not a new phenomenon. In fact, there are several standard techniques used in traditional computing to cope with some of the more common DoS techniques, although this is still an open problem to the network security community. Unfortunately, WSN cannot afford the computational overhead necessary in implementing many of the typical defensive strategies. For example, a sensor network designed to alert building occupants in the event of a fire could be highly susceptible to a DoS attack. DoS on such a WSN could prove very costly, especially on major roads. For this reason, researchers have spent a great deal of time both identifying the various types of DoS attacks and devising strategies to subvert such attacks.

Types of DoS attacks
A standard attack on WSN is simply to jam a node or set of nodes. Jamming, in this case, is simply the transmission of a radio signal that interferes with the radio frequencies being used by the WSN [14]. The jamming of a network can come in two forms: constant jamming (No messages are able to be sent or received.), and intermittent jamming (nodes are able to exchange messages periodically).

Attacks can also be made on the link layer itself. One possibility is that an attacker may simply intentionally violate the communication protocol, e.g., ZigBee [15] or IEEE 801.11 b (Wi-Fi) protocol, and continually transmit messages in an attempt to generate collisions. Such collisions would require the retransmission of any packet affected by the collision. Using this technique it would be possible for an attacker to simply deplete a sensor node's power supply by forcing too many retransmissions.

At the routing layer, a node may take advantage of a multihop network by simply refusing to route messages. This could be done intermittently or constantly with the net result being that any neighbor who routes through the malicious node will be unable to exchange messages with, at least, part of the network. The transport layer is also susceptible to attack, as in the case of flooding. Flooding can be as simple as sending many connection requests to a susceptible node. In this case, resources must be allocated to handle the connection request. Eventually a node's resources will be exhausted, thus rendering the node useless (Table 1.1).

The Sybil attack

Newsome et al. describe the Sybil attack as it relates to WSN [16]. Simply put, the Sybil attack is defined as a "malicious device illegitimately taking on multiple identities" [16]. It was originally described as an attack able to defeat the redundancy mechanisms of distributed data storage systems in peer-to-peer networks [17]. In addition to defeating distributed data storage systems, the Sybil attack is also effective against routing algorithms, data aggregation, voting, fair resource allocation and foiling misbehavior detection. Regardless of the target (voting, routing, aggregation), the Sybil algorithm functions similarly. All of the techniques involve utilizing multiple identities. For instance, in a sensor network voting scheme, the Sybil attack might utilize multiple identities to generate additional "votes." Similarly, to attack the routing protocol, the Sybil attack would rely on a malicious node taking on the identity of multiple nodes, and thus routing multiple paths through a single malicious node.

Table 1.1 Sensor network layers and DoS attacks/defenses

Network layers	Attacks	Defenses
Physical	Jamming	Spread-spectrum, priority messages, lower duty cycle, region mapping, mode change
	Tampering	Tamper-proof, hiding
Link	Collision	Error correcting code
	Exhaustion	Rate limitation
	Unfairness	Small frames
Network and routing	Neglect and greed Homing	Redundancy, probing Encryption
	Misdirection	Egress filtering, authorization monitoring
	Black holes	Authorization, monitoring, redundancy
Transport	Flooding	Client puzzles
	Desynchronization	Authentication

Traffic Analysis Attacks

WSN are typically composed of many low-power sensors communicating with a few relatively robust and powerful base stations. It is not unusual, therefore, for data to be gathered by the individual nodes where it is ultimately routed to the base station. Often, for an adversary to effectively render the network useless, the attacker can simply disable the base station (BS).

A rate monitoring attack simply makes use of the idea that nodes closest to the base station tend to forward more packets than those farther away from the base station. An attacker need only monitor which nodes are sending packets and follow those nodes that are sending the most packets. In a time correlation attack, an adversary simply generates events and monitors to whom a node sends its packets [18].

Node Replication Attacks

Conceptually, a node replication attack is quite simple: an attacker seeks to add a node to an existing WSN by copying (replicating) the node ID of an existing sensor node [19]. A node replicated in this fashion can severely disrupt a sensor network's performance: packets can be corrupted or even misrouted. This can result in a disconnected network, false sensor readings, etc.

Attacks Against Privacy

WSN technology promises a vast increase in automatic data collection capabilities through efficient deployment of tiny sensor devices. While these technologies offer great benefits to users, they also exhibit significant potential for abuse. Particularly relevant concerns are privacy problems, since sensor networks provide increased data collection capabilities [20].

The main privacy problem is not that sensor networks enable the collection of information. In fact, much information from WSN could probably be collected through direct site surveillance. Rather, sensor networks aggravate the privacy problem because they make large volumes of information easily available through remote access. Hence, adversaries need not be physically present to maintain surveillance. They can gather information in a low-risk, anonymous manner. Some of the more common attacks [20, 21] against sensor privacy are:

- Monitor and Eavesdropping. This is the most obvious attack to privacy. By listening to the data, the adversary could easily discover the communication contents.
- Traffic Analysis typically combines with monitoring and eavesdropping. An increase in the number of transmitted packets between certain nodes could signal that a specific sensor has registered activity. Through the analysis on the traffic, some sensors with special roles or activities can be effectively identified.
- Camouflage. Adversaries can insert their node or compromise the nodes to hide in the sensor network. After that these nodes can masquerade as a normal node to attract the packets, then misroute the packets.

Physical Attacks
Sensor networks typically operate in hostile outdoor environments in which the small form factor of the sensors, coupled with the unattended and distributed nature of their deployment make them highly susceptible to physical attacks, i.e., threats due to physical node destructions [22]. Physical attacks destroy sensors permanently, unlike many other attacks mentioned above, so the losses are irreversible. For instance, attackers can extract cryptographic secrets, tamper with the associated circuitry, modify programming in the sensors, or replace them with malicious sensors under the control of the attacker [23].

1.1.4 Defensive Measures

Key Establishment
One security aspect that receives a great deal of attention in WSN is the area of key management. WSN are unique (among other embedded wireless networks) in this aspect due to their size, mobility and computational/power constraints. Indeed, researchers envision WSN to be orders of magnitude larger than their traditional embedded counterparts. This, coupled with the operational constraints described previously, makes secure key management an absolute necessity in most WSN designs. Encryption and key management/establishment are so crucial to the defense of a WSN, with nearly all aspects of WSN defenses relying on solid encryption.

Key Establishment and Associated Protocols
Random key pre-distribution schemes have several variants [6, 24–26]. Eschenauer and Gligor propose a key pre-distribution scheme [6] that relies on probabilistic key sharing among nodes within the sensor network. Their system works by distributing a key ring to each participating node in WSN before deployment. Each key ring should consist of a number randomly chosen keys from a much larger pool of keys generated offline. An enhancement to this technique utilizing multiple keys is described in [27]. Further enhancements are proposed in [26, 28] with additional analysis and enhancements provided by [25].

The LEAP protocol described by Zhu et al. [29] takes an approach that utilizes multiple keying mechanisms. Their observation is that no single security requirement accurately suites all types of communication in a WSN. Therefore, four different keys are used depending on whom the sensor node is communicating with. Sensors are preloaded with an initial key from which further keys can be established.

In PIKE [30], Chan and Perrig describe a mechanism for establishing a key between two sensor nodes that is based on the common trust of a third node somewhere within the sensor network. The nodes and their shared keys are spread over the network such that for any two nodes A and B, there is a node C that shares a key with both A and B.

Public Key Cryptography

Two of the major techniques used to implement public-key cryptosystems are RSA and elliptic curve cryptography (ECC) [31]. In [32] Gura et al. report that both RSA and elliptic curve cryptography are possible using 8-bit CPUs with ECC, demonstrating a performance advantage over RSA. Another advantage is that ECC's 160 bit keys result in shorter messages during transmission compared the 1024 bit RSA keys. In [33], Watro et al. show that portions of the RSA cryptosystem can be successfully applied to actual wireless sensors, specifically the UC Berkeley MICA2 motes [34]. In particular, they implemented the public operations on the sensors themselves while offloading the private operations to devices better suited for the larger computational tasks.

Defending Against DoS Attacks

Since DoS attacks are so common, effective defenses must be available to combat them. One strategy in defending against the classic jamming attack is to identify the jammed part of the sensor network and effectively route around the unavailable portion. Wood and Stankovic [14] describe a two phase approach where the nodes along the perimeter of the jammed region report their status to their neighbors who then collaboratively define the jammed region and simply route around it. To handle jamming at the MAC layer, nodes might utilize a MAC admission control that is rate limiting. This would allow the network to ignore those requests designed to exhaust the power reserves of a node.

To overcome the transport layer flooding DoS attack Aura, Nikander and Leiwo suggest using the client puzzles posed by Juels and Brainard [35] in an effort to discern a node's commitment to making the connection by utilizing some of their own resources. This strategy would likely be effective as long as the client has computational resources comparable to those of the server.

Secure Broadcasting and Multicasting

The research community of WSN has progressively reached a consensus that the major communication pattern of WSN is broadcasting and multicasting, e.g., 1-to-N, N-to-1, and M-to-N, instead of the traditional point-to-point communication on the Internet. In WSN, a great deal of the security derives from ensuring that only members of the broadcast or multicast group possess the required keys in order to decrypt the broadcast or multicast messages.

Traditionally, multicasting and broadcasting techniques have been used to reduce the communication and management overhead of sending a single message to multiple receivers. In order to ensure that only certain users receive the multicast or broadcast, encryption techniques must be employed. In both a wired and wireless network this is done using cryptography. The problem then is one of key management. To handle this, several key management schemes have been devised: centralized group key management protocols, decentralized management protocols, and distributed management protocols [36].

Defending Against Attacks on Routing Protocols
Routing in WSN has, to some extent, been reasonably well studied. However, most current research has focused primarily on providing the most energy efficient routing. There is a great need for both secure and energy efficient routing protocols in WSNs as attacks such as the sinkhole, wormhole and Sybil attacks demonstrate [13, 37].

As WSNs continue to grow in size and utility, routing security must not be an after-thought, but rather they must be included as part of the overall sensor network design. This section describes the current state of routing security as it applies to WSNs.

Defending Against the Sybil Attack
To defend against the Sybil attack the network needs some mechanism to validate that a particular identify is the only identity being held by a given physical node [16]. Newsome et al. primarily describe direct validation techniques, including a radio resource test. In the radio test, a node assigns each of its neighbors a different channel on which to communicate. The node then randomly chooses a channel and listens. Another technique to defend against the Sybil attack is to use random key pre-distribution techniques. The idea behind this technique is that with a limited number of keys on a key-ring, a node that randomly generates identities will not possess enough keys to take on multiple identities and thus will be unable to exchange messages on the network due to the fact that the invalid identity will be unable to encrypt or decrypt messages.

Detecting Node Replication Attacks
In Parno et al. [19], describe two algorithms: randomized multicast, and line-selected multicast. Randomized multicast is an evolution of a node broadcasting strategy. In the simple node broadcasting strategy each sensor propagates an authenticated broadcast message throughout the entire WSN. Any node that receives a conflicting or duplicated claim revokes the conflicting nodes [19]. This strategy will work, but the communication cost is far too expensive. In order to reduce the communication cost, a deterministic multicast could be employed where nodes would share their locations with a set of witness nodes. In this case, witnesses are computed based on a node's ID.

The line-selected multicast algorithm seeks to further reduce the communication costs of the randomized multicast algorithm. It is based upon rumor routing described in [38]. The idea is that a location claim traveling from source s to destination d will also travel through several intermediate nodes. If each of these nodes records the location claim, then the path of the location claim through the network can be thought of as a line segment [19].

Combating Traffic Analysis Attacks
Deng et al. propose using a random walk forwarding technique that occasionally forwards a packet to a node other than the sensor's parent node [18]. This would make it difficult to discern a clear path from the senor to the base station and would help to mitigate the rate monitoring attack, but would still be vulnerable to the time

correlation attack. To defend against the time correlation attack, Deng et al. suggest a fractal propagation strategy [18]. In this technique a node will (with a certain probability) generate a fake packet when its neighbor is forwarding a packet to the base station. The fake packet is sent randomly to another neighbor who may also generate a fake packet. These packets essentially use a time-to-live (TTL) to decide when forwarding should stop. This effectively hides the base station from time correlation attacks.

Defending Against Attacks on Sensor Privacy

Regarding the attacks on privacy mentioned earlier, there are some effective techniques to counter many of the attacks levied against a sensor. Several common techniques are described here [20].

Location information that is too precise can enable the identification of a user, or make the continued tracking of movements feasible. This is a threat to privacy. Anonymity mechanisms depersonalize the data before the data is released, which present an alternative to privacy policy-based access control. Researchers have discussed several approaches using anonymity mechanisms, for example, Gruteser and Grunwald [39] analyze the feasibility of anonymous location information for location-based services in an automotive telematic environment; Beresford and Stajano [40] independently evaluate anonymity techniques for an indoor location system based on the Active Bat.

Policy-based approaches are currently a hot approach to address the privacy problem. The access control decisions and authentication are made based on the specifications of the privacy policies. In [41], Molnar and Wagner present the concept of private authentication, and give a general scheme for building private authentication with work logarithmic in the number of tags in (but not limited by) RFID (radio frequency identification) applications.

Intrusion Detection

Many secure routing schemes attempt to identify network intruders, and key establishment techniques are used in part to prevent intruders from overhearing network data. Despite the necessity of effective intrusion detection schemes for WSNs, a good solution has not yet been devised. Of course, this is due largely to the resource constraints present in WSNs. However, resource constraints are not the only reason. Another problem is that researchers have not yet been able to develop methods of reliably detecting intruders in sensor networks. As such, it is difficult to define characteristics (or signatures) that are specific to a network intrusion as opposed to the normal network traffic that might occur as the result of normal network operations or malfunctions resulting from the environment change.

Secure Data Aggregation

As WSN continue to grow in size, so does the amount of data that the sensor networks are capable of sensing. However, due to the computational constraints placed on individual sensors, a single sensor is typically responsible for only a small part of the overall data. Because of this, a query of WSN is likely to return a great deal of raw data, much of which is not of interest to the individual performing the query. However, such a technique is particularly vulnerable to attacks as a single node is used to aggregate multiple data. Because of this, secure information aggregation techniques are needed in WSNs where one or more nodes may be malicious.

Defending Against Physical Attacks
Physical attacks pose a great threat to WSN, because of its unattended feature and limited resources. Sensor nodes may be equipped with physical hardware to enhance protection against various attacks. For example, to protect against tampering with the sensors, one defense involves tamper-proofing the node's physical package [14]. Another way is to employ special software and hardware outside the sensor to detect physical tampering. As the price of the hardware itself gets cheaper, tamper-resistant hardware may become more appropriate in a variety of sensor network deployments.

One possible approach to protect the sensors from physical attacks is self-termination. The basic idea is the sensor kills itself, including destroy all data and keys, when it senses a possible attack. This is particularly feasible in the large scale WSN which has enough redundancy of information, and the cost of a sensor is much cheaper than the lost of being broken (attacked).

Trust Management
Trust is an old but important issue in any networked environment, whether in social networking or in computer networking. Trust can solve some problems beyond the power of the traditional cryptographic security. For example, judging the quality of the sensor nodes and the quality of their services, and providing the corresponding access control. The trust issue is emerging as sensor networks thrive. However, it is not easy to build a good trust model within WSN given the resource limits. Zhu et al. [42] provide a practical approach to compute trust in WSN by viewing individual mobile devices as a node of a delegation graph G and mapping a delegation path from the source node S to the target node T into an edge in the correspondent transitive closure of the graph G, from which the trust value is computed. In this approach, an undirected transitive signature scheme is used within the authenticated transitive graphs.

1.2 Attackers Motivation

Motivation refers to the benefit the attacker hopes to gain from the attack. This can be further broken down into one of two classes of gains:

- **Benefit from Data**: One of the motivations for attacking a WSN for some applications is to gain access to the sensitive data being monitored or relayed. Thus, the goal of the attacker is access to the data being carried or meta data about the users or their activity. The emphasis for these types of applications is on confidentiality and privacy preserving measures.

- **Mission Interference**: Another motivation for attacking a WSN is to interfere with its mission. In this case, the data carried by the WSN is not necessarily of interest to the attacker, who instead desires to compromise the WSN's ability to function. In these types of applications, the adversary is often being monitored and desires to circumvent this monitoring by falsifying data or disrupting the network or a subset of it. Here, attacks on the infrastructure and services enabling the WSN, or attacks allowing tampering with the data can achieve the desired effect. Note that not all points in the WSN are of equal benefit for disruption: Disrupting critical relay nodes, nodes with unique coverage or even the base station can result in disproportionately more damage than some redundant sensor that does not play an important role. Further, we distinguish between attacks that are detectable, and those that are not. In the latter case, the attacker's benefit may be enhanced because the observer acts based on bad or manipulated data. If the failure is detectable, the observer may employ backup monitoring mechanisms or ignore the WSN as a valid source of data.

These two types of benefits may exist concurrently in an application. Further, in sensor and actuator networks, the benefit may be in terms of the action taken (or not taken) by the actuators. Regardless of the mode of benefit, the relative degree of benefit is an indicator of the motivation of an attacker to attack as well as their relative preference among the different attacks. Thus, it is also an indicator of how much the designer and operator of the WSN should protect against these attacks. Finally, we note that accurately quantifying benefit is difficult; often human estimates are used for utility in similar contexts. Finally, some attacks such as vandalism may occur that have no tangible benefit to attackers.

1.3 Research Challenges

In this brief chapter, I describe six key research challenges for WSN [43].

Security in Real-world Protocols

Many current WSN solutions are developed with simplifying assumptions about wireless communication and the environment, even though the realities of wireless communication and environmental sensing are well known. Many of these solutions work very well in simulation. It is either unknown how the solutions work in the real world or they can be shown to work poorly in practice. We note that, in general, there is an excellent understanding of both the theoretical and practical issues related to wireless communication. For example, it is well known how the signal strength drops over distance. Effects of signal reflection, scattering and fading are understood. However, when building an actual WSN, many specific system, application, and cost issues also affect the communication properties of the system. The size, power, cost constraints and their tradeoffs are fundamental constraints. In the current state of the art, the tradeoff among these constraints has produced a number of devices currently being used in WSNs. For example, one such device is the Mica mote that uses 2 AA batteries, a 7 MHz microcontroller, an RF Chipcon radio, and costs about $100. As better batteries, radios, and microcontrollers become available and as costs reduce, new platforms will be developed. These new platforms will continue to have tradeoffs between these parameters. Novel network protocols that account for the key realities in wireless communication are required.

Security in Real-Time

WSN deals with real world environments. In many cases, sensor data must be delivered within time constraints so that appropriate observations can be made or actions taken. Very few results exist to date regarding meeting real-time requirements in WSN. Most protocols either ignore real-time or simply attempt to process as fast as possible and hope that this speed is sufficient to meet deadlines. Some initial results exist for real-time routing. For example, the RAP protocol [44] proposes a new policy called velocity monotonic scheduling. Here a packet has a deadline and a distance to travel. Using these parameters a packet's average velocity requirement is computed and at each hop packets are scheduled for transmission based on the highest velocity requirement of any packets at this node. While this protocol addresses real-time, no guarantees are given. Another routing protocol that addresses real-time is called SPEED [45]. This protocol uses feedback control to guarantee that each node maintains an average delay for packets transiting a node. Transient behavior, message losses, congestion, noise and other problems cause these guarantees to be limited. To date, the limited results that have appeared for WSN regarding real-time issues has been in routing. Many other functions must also meet real-time constraints including: data fusion, data transmission, target and event detection and classification, query processing, and security. New results are needed to guarantee soft real-time requirements and that deal with the realities of WSN such as lost messages, noise and congestion.

Power Management

Low-cost deployment is one acclaimed advantage of sensor networks. Limited processor bandwidth and small memory are two arguable constraints in sensor networks, which will disappear with the development of fabrication techniques. However, the energy constraint is unlikely to be solved soon due to slow progress in developing battery capacity. Moreover, the untended nature of sensor nodes and hazardous sensing environments preclude battery replacement as a feasible solution. On the other hand, the surveillance nature of many sensor network applications requires a long lifetime; therefore, it is a very important research issue to provide a form of energy-efficient surveillance service for a geographic area. Based on the fact that individual sensor nodes are not reliable and subject to failure and single sensing readings can be easily distorted by background noise and cause false alarms, it is simply not sufficient to rely on a single sensor to safeguard a critical area. In this case, it is desired to provide higher degree of coverage in which multiple sensors monitor the same location at the same time in order to obtain high confidence in detection. On the other hand, it is overkill and energy consuming to support the same high degree of coverage for some non-critical area.

Programming Abstractions
A key to the growth of WSN is raising the level of abstraction for programmers. Currently, programmers deal with too many low levels details regarding sensing and node to node communication. For example, they typically deal with sensing data, fusing data and moving data. They deal with particular node to node communication and details. If we raise the level of abstraction to consider aggregate behavior, application functionality and direct support for scaling issues then productivity increases. Current research in programming abstractions for WSN can be categorized into 7 areas: environmental, middleware APIs, database centric, event based, virtual machines, scripts and component-based.

Security and Privacy
WSN are limited in their energy, computation, and communication capabilities. In contrast to traditional networks, sensor nodes are often deployed in accessible areas, presenting a risk of physical attacks. WSN interact closely with their physical environment and with people, posing additional security problems. Because of these reasons current security mechanisms are inadequate for WSN. These new constraints pose new research challenges on key establishment, secrecy and authentication, privacy, robustness to denial-of-service attacks, secure routing, and node capture. To achieve a secure system, security must be integrated into every component, since components designed without security can become a point of attack. Consequently, security and privacy pervade every aspect of system design.

Analysis
Few analytical results exist for WSN. Since WSN are in the early stage of development it is not surprising that few analytical results exist. Researchers are busy inventing new protocols and new applications for WSN. The solutions are built, tested and evaluated either by simulation or testbeds; sometimes an actual system has been deployed. Empirical evidence is beginning to accumulate. However, a more scientific approach is required where a system can be designed and analyzed before it is deployed. The analysis needs to provide confidence that the system will meet its requirements and to indicate the efficiency and performance of the system.

Summary
In this brief note six key research areas were highlighted. However, many other research areas are very important including: localization, topology control, dependability, self-calibration, self-healing, data aggregation, group management, clock synchronization, query processing, sensor processing and fusion under limited capacities, and testing and debugging.

1.4 Motivation of the Book

The previous list of attacks and research challenges allow us to formulate the research questions investigated in this book. The questions are:

1. What's the main purpose of WSN security? Is security need for WSN? What are the key security challenges?

WSN is exposed to new trend of attacks, although a lots of countermeasures methods have been extensively studied to provide WSN communication securities. These defenses are ineffective against attacks i.e. from compromised servers due to WSN level constantly increasing, attacks and becoming more and more complicated. And as information become more valuable and costly making intruders to use more complicated methods in attacking WSN, eventually this makes security issue become highly sensitive.

2. Is WSN need new models of security? Why? How QoS can be used for achieve WSN security?

WSNs are increasingly being used in applications where QoS and low cost are the overriding considerations. Conventional schemes of using sensor nodes and incorporating three security phenomenons (reliability, availability and serviceability) to attain QoS can effectively improve not only the reliability of the overall WSNs but security as well. There are some works discussed QoS problems in WSN, but they have not focused on the availability, reliability and serviceability together as means of providing security integrity in WSN. The reverse look of QoS can be used and hence present mathematically the three significant quality factors that should be currently taken into account in developing WSN application services and security. QoS factors required when developing security applications and QoS can be integrated to increase the security and applications running in WSNs.

3. Is WSN need new mathematical models for security? Why this method should combine internal and external security mechanisms?

Existing sensor network security research has mostly focused on adapting security mechanisms to the computational and messaging constraints imposed by tiny sensor devices. Efforts are currently underway to extend the scalability of WSNs so that they can be used to monitor one of the largest international borders. Intrusion detection and border surveillance constitute a major application category for WSNs. WSNs have recently emerged as an important means to study and interact with physical world. So to combine internal and external security mechanisms, the new mathematical model can be proposed. For WSN security, the theoretical foundations for laying barriers of wireless sensors can be developing. Fundamental disk models and mathematical models can be designed which provide basic building blocks to implement various security mechanisms for sensor nodes security. These models should be flexible to be embedded in sensor nodes and should create a suitable nodes components security in hostile environments.

4. Are improved ciphers and protocols needed to WSN?

Efficient and resources constrained security mechanisms for WSN is needed, new and more stable security approaches need to be put in place to provide information safety considering the following attributes: availability, confidentiality, integrity, authentication, and non-repudiation. A modified accelerated-cipher for WSN can be used for fast hardware, firmware, software and WSN encryption systems. The approach should provide ciphers using this approach have less intrusion probability against differential cryptanalysis than currently used popular WSN ciphers.

5. Why the new signature distributed schemes are need? How are they providing better security in WSN?

Signature distributed scheme is important security service, because the problem is to establish a secure signature schemes between communicating nodes (it enables sensor nodes to communicate securely with each other). So, special and reliable aggregation algorithms when it comes to node failing (compromised) fulfilling their tasks are needed. Distributed signature scheme can be designed for solving this problem. This scheme should have three advantages: the mathematics presentations are provably secure, the scheme is efficient, and efficient proactive scheme with three security properties which did not exist in previously schemes.

6. Why new reliable data aggregation protocol is need to WSN? How it can protect aggregation of data and ensure WSN reliability?

Current routing protocols in WSNs or even in Wireless Ad hoc Networks are very susceptible to many attack i.e. stealthy attack. To answer this question, first, we abstract two fundamental routing protocols, which can be generally grouped in two broad categories based on the intrinsic nature of WSN. We argue that none of previous proposed routing protocols satisfies all of them at the same time. The novelty of new protocol is building a general routing protocol based on two methods, which takes into consideration two factors, first additional of sensor nodes to the aggregation process and second by considering complex report interaction between base station and aggregator.

Security is an overhead to the existing network QoS measurements therefore it has a strong influence on QoS of a network. QoS metrics such as authentication delay, mobility, cost, call dropping probability and throughput of communication due to authentication overhead has to be affected.

Size and number of packets transmitted is increased to include security parameters which affect the payload of messages. To analyze it we need to study the effects of QoS with security in place, so we thoroughly detailed security metrics (in Chap. 2) and then found connection and transmitting time for different number of packets.

Typically authentication delay causes a pause for data transmission which decreases the throughput. We investigated that effect through new accelerated ciphers design in Chap. 4, through distributed signature scheme establishment in Chap. 5 and through designing the reliable data aggregation protocol in Chap. 6.

Moreover length of keys and complexity of algorithm used also has an adverse effect which has been researched in Chaps. 4 and 5.

1.5 Contribution of the Book

This book studied the following problems:

- QoS as tools to achieve better WSN security
- Mathematical models for WSN security
- Problem of improved ciphers development for WSN
- Problem of establishment of distributed signature for data transferring
- Problem of reliable data aggregation in WSN

The main contributions include:

- Investigation of new attacks trends, research challenges and possible solutions for WSN security
- QoS as means towards WSN security
- New mathematical model for WSN security
- An improved cipher based on Feistel scheme to design more accelerated ciphers to WSN
- Improvement of distributed signature scheme for WSN
- New reliable data aggregation protocol for WSN

1.6 Outline of the Book

Chapter 2: *QoS as means of providing WSN security* research how the QoS can be used to achieve WSN security and why WSN needs the new security model.

Chapter 3: *Mathematical model for WSN nodes security* consider why new mathematical models for WSN security should combine internal and external security mechanisms.

Chapter 4. *An improved Feistel-based cipher for WSN security* examines the attacks trends, research challenges and possible solutions through the implementations of the new block ciphers.

Chapter 5: *The distributed signature scheme based on RSA* investigate the necessity of the new signature scheme and answer the question how it can provide better security in WSN.

Chapter 6: *Reliable data aggregation protocol for WSN* examines the well-known data aggregation protocols for WSN and proposes the new secured protocol which can help to achieve the better WSN protection (Fig. 1.2).

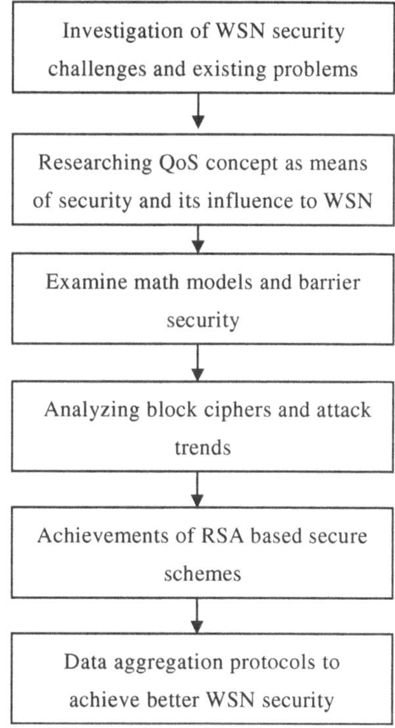

Fig. 1.2 Sequence of study and methodology

References

1. Walters, J.P., Liang, Z.: Wireless sensor network security: a survey. In: Xiao, Y. (eds.) Chapter 17: security in distributed grid, and pervasive computing, Auerbach Publications, CRC Press (2006)
2. Carman D.W., Krus, P.S., Matt, B.J.: Constraints and approaches for distributed sensor network security. Technical Report 00-010, NAI Labs, Network Associates, Inc., Glenwood (2000)
3. Akyildiz, I.F., Su, W., Sankarasubramaniam, Y., Cayirci, E.: A survey on sensor networks. IEEE Commun. Mag. **40**(8), 102–114 (2002)
4. Stankovic, J.A., et al.: Real-time communication and coordination in embedded sensor networks. Proc. IEEE **91**(7), 1002–1022 (2003)
5. Perrig, A., Szewczyk, R., Tygar, J.D., Wen, V., Culler, D.E.: Spins: security protocols for sensor networks. Wireless Networking **8**(5), 521–534 (2002)
6. Eschenauer, L., Gligor, V.D.: A key-management scheme for distributed sensor networks. In: Proceedings of the 9th ACM Conference on Computer and Communications Security, pp. 41–47 (2002)
7. Ganeriwal, S., Capkun, S., Han, C.C., Srivastava, M.B.: Secure time synchronization service for sensor networks. In: WiSe '05: Proceedings of the 4th ACM Workshop on Wireless Security, pp. 97–106 (2005)
8. Capkun, S., Hubaux, J.-P.: Secure positioning in wireless networks. IEEE J. Sel. Areas Commun. **24**(2), 221–232 (2006)
9. Lazos, L., Poovendran, R.: Serloc: robust localization for wireless sensor networks. ACM Trans. Sen. Netw. **1**(1), 73–100 (2005)

10. Liu, D., Ning, P.: Efficient distribution of key chain commitments for broadcast authentication in distributed sensor networks. In: Proceedings of the 10th Annual Network and Distributed System Security Symposium, pp. 263–276 (2003)
11. Liu, D., Ning, P.: Multilevel μTESLA: broadcast authentication for distributed sensor networks. Trans. Embedded Comput. Syst. **3**(4), 800–836 (2004)
12. Perrig, A., Stankovic, J., Wagner, D.: Security in wireless sensor networks. Commun. ACM **47**(6), 53–57 (2004)
13. Karlof, C., Wagner, D.: Secure routing in wireless sensor networks: attacks and countermeasures. Elsevier's AdHoc Netw. J. Spec. Issue Sens. Netw. Appl. Protoc. **1**(2–3), 293–315 (2003)
14. Wood, A.D., Stankovic, J.A.: Denial of service in sensor networks. Computer **35**(10), 54–62 (2002)
15. http://www.zigbee.org/, (2005)
16. Newsome, J., Shi, E., Song, D., Perrig, A.: The Sybil attack in sensor networks: analysis & defenses. In: Proceedings of the Third International Symposium on Information Processing in Sensor Networks, pp. 259–268 (2004)
17. Douceur, J.: The Sybil attack. In: Proceedings of the 1st International Workshop on Peer-to-peer Systems (IPTPS'02), pp. 251–260 (2002)
18. Deng, J., Han, R., Mishra, S.: Countermeasuers against traffic analysis in wireless sensor networks. Technical Report CU-CS-987-04, University of Colorado at Boulder (2004)
19. Parno, B., Perrig, A., Gligor, V.: Distributed detection of node replication attacks in sensor networks. In: Proceedings of IEEE Symposium on Security and Privacy, pp. 49–63 (2005)
20. Gruteser, M., Schelle, G., Jain, A., Han, R., Grunwald, D.: Privacy-aware location sensor networks. In: 9th USENIX Workshop on Hot Topics in Operating Systems (HotOS IX), vol. 9, p. 28 (2003)
21. Chan, H., Perrig, A.: Security and privacy in sensor networks. IEEE Comput. Mag., 103–105 (2003)
22. Wang, X., Gu, W., Schosek, K., Chellappan, S., Xuan, D.: Sensor network configuration under physical attacks. Technical Report Technical Report (OSU-CISRC-7/04-TR45), Department of Computer Science and Engineering, The Ohio-State University (2004)
23. Wang, X., Gu, W., Chellappan, S., Xuan, D., Ten H. Laii.: Search-based physical attacks in sensor networks: modeling and defense. Technical report, Department of Computer Science and Engineering, The Ohio-State University, 2005
24. Chan, H., Perrig, A., Song, D.: Random key predistribution schemes for sensor networks. In: Proceedings of the 2003 IEEE Symposium on Security and Privacy, p. 197 (2003)
25. Hwang, J., Kim, Y.: Revisiting random key pre-distribution schemes for wireless sensor networks. In: Proceedings of the 2nd ACM Workshop on Security of Ad hoc and Sensor Networks (SASN '04), pp. 43–52 (2004)
26. Liu, D., Ning, P., Li, R.: Establishing pairwise keys in distributed sensor networks. ACM Trans. Inf. Syst. Secur. **8**(1), 41–77 (2005)
27. Blakely, G.R.: Safeguarding cryptographic keys. In: Proceedings of the National Computer Conference, vol. 48 (1979)
28. Du, W., Deng, J., Han, Y.S., Varshney, P.K.: A pairwise key pre-distribution scheme for wireless sensor networks. In: Proceedings of the 10th ACM Conference on Computer and Communications Security, pp. 42–51 (2003)
29. Zhu, S., Setia, S., Jajodia, S.: Leap: efficient security mechanisms for largescale distributed sensor networks. In: Proceedings of the 10th ACM Conference on Computer and Communications Security, pp. 62–72 (2003)
30. Chan, H., Perrig, A.: Pike: peer intermediaries for key establishment in sensor networks. IEEE Infocom. **1**, 524–535 (2005)
31. Schneier, B.: Applied cryptography: protocols, algorithms, and source code (Second edition). John Wiley and Sons, New York, p. 758 (1996)

32. Gura, N., Patel, A., Wander, A., Eberle, H., Shantz, S.: Comparing elliptic curve cryptography and RSA on 8-bit cpus. In: 2004 Workshop on Cryptographic Hardware and Embedded Systems, pp. 119–132 (2004)
33. Watro, R., Kong, D., Cuti, S., Gardiner, C., Lynn, C., Kruus, P.: Tinypk: securing sensor networks with public key technology. In: Proceedings of the 2nd ACM Workshop on Security of Ad hoc and Sensor Networks (SASN '04), pp. 59-64 (2004)
34. Hill, J., Szewczyk, R., Woo, A., Hollar, S., Culler, D.E., Pister, K.: System architecture directions for networked sensors. In: Architectural Support for Programming Languages and Operating Systems, pp. 93–104 (2000)
35. Aura, T., Nikander, P., Leiwo, J.: Dos-resistant authentication with client puzzles. Lecture Notes in Computer Science **2133**: 170–177 (2001)
36. Rafaeli, S., Hutchison, D.: A survey of key management for secure group communication. ACM Comput. Surv. **35**(3), 309–329 (2003)
37. Hu, Y., Perrig, A., Johnson, D.B.: Packet leashes: a defense against wormhole attacks in wireless networks. In: INFOCOM 2003. Twenty-Second Annual Joint Conference of the IEEE Computer and Communications Societies, vol. 3, pp. 1976–1986 (2003)
38. Braginsky, D., Estrin, D.: Rumor routing algorthim for sensor networks. In: Proceedings of the 1st ACM International Workshop on Wireless Sensor Networks and Applications, pp. 22–31 (2002)
39. Gruteser, M., Grunwald, D.: Anonymous usage of location-based services through spatial and temporal cloaking. In: Proceedings of the First International Conference on Mobile Systems, Applications, and Services (MobiSys). USENIX, pp. 31–42 (2003)
40. Beresford, A.R., Stajano, F.: Location privacy in pervasive computing. IEEE Pervasive Comput. **2**(1), 46–55 (2003)
41. Molnar, D., Wagner, D.: Privacy and security in library RFID. Issues, practices, and architectures. In: Proceedings of the 11th ACM Conference on Computer and Communications Security, pp. 210–219 (2004)
42. Zhu, H., Bao, F., Deng, R.H., Kim, K.: Computing of trust in wireless networks. In: Proceedings of 60th IEEE Vehicular Technology Conference, vol. 4, pp. 2621–2624 (2004)
43. Stankovic, John A.: Research challenges for wireless sensor networks. ACM SIGBED Rev. **1**, 9–12 (2004)
44. Lu, C., Blum, B., Abdelzaher, T., Stankovic, J., He, T.: RAP: a real-time communication architecture for large-scale wireless sensor networks, Technical Report: CS-2002-10 (2002)
45. He, T., Stankovic, J., Lu, C., Abdelzaher, T.: SPEED: a stateless protocol for real-time communication in Ad Hoc sensor networks. In Proceedings of the 23rd International Conference on Distributed Computing Systems, p. 46 (2003)

Chapter 2
QoS as Means of Providing WSN Security

2.1 Introduction

WSN are being increasingly used in applications where QoS and low cost are the overriding considerations. With increased use, their reliability, availability and serviceability need to be addressed from the outset. Conventional schemes of using sensor nodes and incorporating these three phenomenons (reliability, availability and serviceability—RAS) to attain QoS can effectively improve not only the reliability of the overall WSNs but security as well. We discuss the reverse look of QoS and hence present mathematically the three significant quality factors that should currently be taken into account in developing WSNs application services and security availability, reliability and serviceability. We also discussed specific characteristics and constraints of WSN, QoS factors when developing security applications for such networks. The security of WSNs has been addressed by providing the flow models and simulations testing using individual sensor nodes on our experiment. To evaluated possibility of establishing secure WSN through QoS we have used Hawk nodes to demonstrate our approach experimentally. The flow models show how the QoS can be integrated to increases the security of applications running under WSNs.

Sensor network communications must prevent disclosure and undetected modification of exchanged messages. Due to the fact that individual sensor nodes are anonymous and that communication among sensors is via wireless links, sensor networks are highly vulnerable to security attacks. If an adversary can thwart the work of the network by perturbing the information produced, stopping production, or pilfering information, then the perceived usefulness of sensor networks will be drastically curtailed. Thus, security is a major issue that must be resolved in order for the potential of WSNs to be fully exploited.

Several researchers have discussed QoS problems in WSN, but they have not focused on the availability, reliability and serviceability together as means of providing security integrity in WSN.

© Springer International Publishing Switzerland 2016
G.S. Oreku and T. Pazynyuk, *Security in Wireless Sensor Networks*,
Risk Engineering, DOI 10.1007/978-3-319-21269-2_2

QoS is an overused term with multiple meanings and perspectives from different research and technical communities [1]. Perillo et al. [2] have defined QoS as measurements of application reliability with a goal of energy efficiency. An alternative definition equates QoS to spatial resolution [3]. This latter work also presented a QoS control strategy based on a Gur game paradigm in which base stations broadcast feedback to the network's sensors. The former work [2] refers to QoS parameters specific to the application, such as sensor node measurement, deployment, and coverage and number of active sensor nodes. The latter refers to how the supporting communication network can meet application needs while efficiently using network resources such as bandwidth and power consumption.

As WSNs are expected to be adopted in many industrial, health care and military applications, their reliability, availability and serviceability (RAS) are becoming critical. In recent years, the diverse potential applications for WSNs have been touted by researchers [4, 5] and the general press [6]. In many WSNs systems, to provide sufficient RAS can often be absorbed in the network cost. Nevertheless, as noticed early [7], network designers face "two fundamentally conflicting goals: to minimize the total cost of the network and to provide redundancy as a protection against major service interruptions".

To the best of our knowledge our approach is different from most of the existing works and on going research which deal with WSNs strategy to achieve QoS, we extend our finding to QoS as a support to security model we have designed by introducing a number of active security requirements distributed in a gradient fashion based on their logical connection to the QoS requirements.

Traditional QoS mechanisms used in wired networks aren't adequate to support WSNs because of constraints such as resource limitations and dynamic topology. So we build the middleware that provide new mechanisms to maintain QoS over an extended period and even adjust itself when the required QoS and the state of the application changes. That middleware had been designed based on trade-offs among performance metrics such as network capacity or throughput, data delivery delay, and energy consumption. We have considered QoS with new point of view, i.e. QoS as means of providing security mechanism in WSN. From combined availability, reliability and serviceability volumes together, we estimated meaning of QoS for WSN and then analyze it through our models.

WSNs are prone to security problems such as the compromising, tampering and malicious intrusions, eavesdropping of sensor data and adversarial packet injection, DoS attacks [8]. With this awareness in mind, integration of Security analysis with QoS design needs to meet both defense from these attacks and satisfy certain QoS requirements simultaneously from long view.

We examine WSNs nodes and propose the necessary QoS required for increasing both the availability and serviceability of the system, also as a way of intensify security within WSNs. Conventional schemes of using sensor nodes and incorporating these three phenomenons (reliability, availability and serviceability) to attain QoS can effectively improve not only the reliability of the overall WSNs

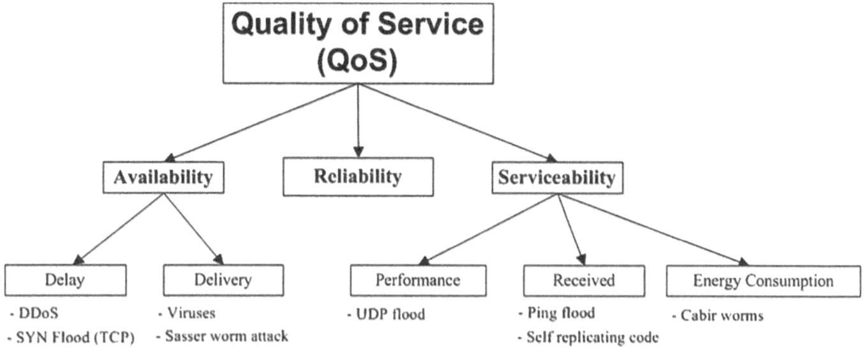

Fig. 2.1 Integration of security analysis with QoS

but also security. We present specific characteristics and constraints of WSNs QoS factors when developing security applications for such networks. The flow models show how the QoS can be integrated to increase the security of applications running under WSNs.

For better understanding, our approach is clearly mapped to our model where the model is merging to QoS as Fig. 2.1 indicates. A secure model is proposed with flow of security classes providing different levels of security using QoS. We assume that security context is not a precise technical specification; rather, it is a set of security-related factors narrowing down the WSNs security design to the region. Evidently, we describe the security context in terms of three related groups of factors and WSNs application: (1) Availability Motivation; (2) Reliability and (3) Serviceability.

2.2 QoS in Wireless Networks

2.2.1 QoS Concept

Quality-of-Service is "a set of service requirements to be met by the network while transporting a flow" [9]. Here a flow is "a packet stream from source to a destination (unicast or multicast) with an associated QoS" [9]. In other words, QoS is a measurable level of service delivered to network users, which can be characterized by packet loss probability, available bandwidth, end-to-end delay, etc. Such QoS can be provided by network service providers in terms of some agreement (Service Level Agreement, or SLA) between network users and service providers. For example, users can require that for some traffic flows, the network should choose a path with minimum 2 Mbit/s bandwidth.

2.2.2 QoS Metrics

To be implemented, service requirements have to be expressed in some measurable QoS metrics. Well-known metrics include bandwidth, delay, jitter, cost, loss probability, etc. Different metrics may have different features. The most commonly used functional forms of QoS-metrics are additive, multiplicative, and concave [10], classified according to an arithmetic relationship between the associated path-metric and link-metric. They are defined as follows.

For a path $P = (n_1, n_2, \ldots, n_n)$ of network nodes n_i, $i = 1, 2, \ldots, n$, a metric m *is*:

1. ***Additive***, if $m(P) = m(n_1, n_2) + m(n_2, n_3) + \cdots + m(n_{n-1}, n_n)$

Examples are delay, jitter, cost and hop-count. For instance, the delay of a path is the sum of the delay of every hop.

2. ***Multiplicative***, if $m(P) = m(n_1, n_2) \times m(n_2, n_3) \times \cdots \times m(n_{n-1}, n_n)$

Example is reliability and loss probability.

3. ***Concave***, if $m(P) = \min\{m(n_1, n_2), m(n_2, n_3), \ldots, m(n_{n-1}, n_n)\}$

Example is bandwidth, which means that the bandwidth of a path is determined by the link with the minimum available bandwidth.

2.2.3 Security and QoS

WSN is only as good as the information it produces. In this respect, the most important concern is information security. Indeed, in most application domains sensor networks will constitute a mission critical component requiring commensurate security protection. Sensor network communications must prevent disclosure and undetected modification of exchanged messages. Due to the fact that individual sensor nodes are anonymous and that communication among sensors is via wireless links, sensor networks are highly vulnerable to security attacks.

Being widely deployed in domains that involve sensitive information, for example, healthcare and rescue; the untethered and large deployment of WSNs in harsh environments increases their exposure to malicious intrusions and attacks such as DoS [8]. In addition, the wireless medium facilitates eavesdropping and adversarial packet injection to compromise the network's functioning. All these factors make security extremely important. Furthermore, sensor nodes have limited power and processing resources, so standard security mechanisms, which are heavy in weight and resource consumption, are unsuitable. These challenges increase the need to develop comprehensive and secure solutions that achieve wider protection, while maintaining desirable network performance. Middleware efforts should concentrate on developing and integrating security in the initial phases of software design, hence achieving different security requirements such as confidentiality, authentication, integrity, freshness, and availability.

2.2.4 QoS Challenges in Sensor Networks

Different from IP network, Sensor network naturally supports multiple service types, thus provides different QoS. The service types range from CBR (Constant Bit Rate) which guarantees bandwidth, delay and delay jitter, to UBR (Unspecified Bit Rate) which virtually provides no guarantees (just like today's "best-effort" IP network). While sensor networks inherit most of the QoS issues from the general wireless networks, their characteristics pose unique challenges. The following are an outline of design considerations for handling QoS traffic in WSNs.

Bandwidth limitation: A typical issue for general wireless networks is securing the bandwidth needed for achieving the required QoS. Bandwidth limitation is going to be a more pressing issue for WSN. Traffic in sensor networks can be burst with a mixture of real-time and non-real-time traffic. Dedicating available bandwidth solely to QoS traffic will not be acceptable. A trade-off in image/video quality may be necessary to accommodate non-real-time traffic. In addition, simultaneously using multiple independent routes will be sometime needed to split the traffic and allow for meeting the QoS requirements. Setting up independent routes for the same flow can be very complex and challenging in sensor networks due energy constraints, limited computational resources and potential increase in collisions among the transmission of sensors.

Removal of redundancy: Sensor networks are characterized with high redundancy in the generated data. For unconstrained traffic, elimination of redundant data messages is somewhat easy since simple aggregation functions would suffice. However, conducting data aggregation for QoS traffic is much more complex. Comparison of images and video streams is not computationally trivial and can consume significant energy resources. A combination of system and sensor level rules would be necessary to make aggregation of QoS data computationally feasible. For example, data aggregation of imaging data can be selectively performed for traffic generated by sensors pointing to same direction since the images may be very similar. Another factor of consideration is the amount of QoS traffic at a particular moment. For low traffic it may be more efficient to cease data aggregation since the overhead would become dominant. Despite the complexity of data aggregation of imaging and video data, it can be very rewarding from a network performance point-of-view given the size of the data and the frequency of the transmission.

Energy and delay trade-off: Since the transmission power of radio is proportional to the distance squared or even higher order in noisy environments or in the non-flat terrain, the use of multi-hop routing is almost a standard in WSNs. Although the increase in the number of hops dramatically reduces the energy consumed for data collection, the accumulative packet delay magnifies. Since packet queuing delay dominates its propagation delay, the increase in the number of hops can, not only slow down packet delivery but also complicate the analysis and the handling of delay-constrained traffic. Therefore, it is expected that QoS routing of sensor data would have to sacrifice energy efficiency to meet delivery requirements. In addition, redundant routing of data may be unavoidable to cope with the typical high error rate in wireless communication, further complicating the trade-off between energy consumption and delay of packet delivery.

Buffer size limitation: Sensor nodes are usually constrained in processing and storage capabilities. Multi-hop routing relies on intermediate relaying nodes for storing incoming packets for forwarding to the next hop. While a small buffer size can conceivably suffice, buffering of multiple packets has some advantages in WSNs. First, the transition of the radio circuitry between transmission and reception modes consumes considerable energy [11] and thus it is advantageous to receive many packets prior to forwarding them. In addition, data aggregation and fusion involves multiple packets. Multi-hop routing of QoS data would typically require long sessions and buffering of even larger data, especially when the delay jitter is of interest. The buffer size limitation will increase the delay variation that packets incur while traveling on different routes and even on the same route. Such an issue will complicate medium access scheduling and make it difficult to meet QoS requirements.

Support of multiple traffic types: Inclusion of heterogeneous set of sensors raises multiple technical issues related to data routing. For instance, some applications might require a diverse mixture of sensors for monitoring temperature, pressure and humidity of the surrounding environment, detecting motion via acoustic signatures and capturing the image or video tracking of moving objects. These special sensors are either deployed independently or the functionality can be included on the normal sensors to be used on demand. Reading generated from these sensors can be at different rates, subject to diverse QoS constraints and following multiple data delivery models, as explained earlier. Therefore, such a heterogeneous environment makes data routing more challenging.

2.3 Effect of Security on QoS

Security services provide information secrecy, data integrity and resource availability for users. Information secrecy means to prevent the improper disclosure of information in the communications, while data integrity is to prevent improper modification of data and resource availability is considered to preventing improper DoS [12, 13].

All these attacks are aiming at one or more [14–16]. Although, security concerns in mobile traditional networks apply to sensor networks, the solutions are not the same. Sensor nodes are tightly constrained in terms of energy, processing, and storage capacities. Once deployed, it is often very difficult to change or recharge batteries for such nodes. This constraint limits the number of conventional techniques that can efficiently be adapted to sensor networks. Wireless communication makes information more vulnerable to attacks. Sensor nodes placed into the physical environments; therefore it is often easy to compromise by an attacker.

In addition, it is effortless to capture them physically and ruin them. However sensors networks composed of heterogeneous nodes with different capabilities.

Security is an overhead to the existing network QoS measurements therefore it has a strong influence on QoS of a network as well as providing (RAS). QoS metrics such as authentication delay, mobility, cost, call dropping probability and throughput of communication due to authentication overhead has to be affected. Typically authentication delay causes a pause for data transmission which decreases the throughput. Moreover length of keys and complexity of algorithm used has an adverse effect. Also size of packets transmitted is increased to include security parameters which affect the payload of messages.

Identifying the possible threats that may face sensor networks will help in designing secure WSNs as these threats are the ones hindering QoS. However in case of a WSNs longer keys would have a disastrous effect on the QoS of the network therefore it is important to classify security levels based on information secrecy, data integrity and resource availability. These aspects can be design into the system with variations of security strength classes.

Security classes indicate the level of protection provided by the QoS for analysis of security.

Class 1, No Authentication: Since no encryption is applied therefore secrecy of data and resource protection is not provided.

Class 2, MAC verification only: No encryption is applied therefore secrecy and resource protection is not provided. However this class provides slight bit of security by MAC authentication whereas a MAC address can be easily hacked.

Class 3, Encrypted Challenge/Response without keys: Encryption is applied only to verify user identity, therefore only legitimate users are allowed to have access to a resource. However since data transmission is not encrypted therefore there is a chance of data being compromised.

Class 4, Encrypted transmission with K-length keys: This class provides the highest level of security. However the length of the key could increase the overall authentication cost in terms of processing time. Higher security level is achieved by using complex cryptographic techniques which involve operations that increase the overhead of transmission and affects the QoS parameters such as authentication cost, authentication delay and packet dropping probability.

We investigated the problems of designing the accelerate block ciphers in Chap. 4 and establishment of distributed signature scheme in Chap. 5. The problem of reliable data transferring was considered in Chap. 6. We made a research about the influences of probabilities of stable system working and system breaking on QoS in Chap. 3.

Effects of security metrics place a lot of burden on the QoS of the overall system thus decreasing performance. Commonly used WSN protocols cannot be used for many reasons. Therefore a new secure transmission protocol, which has been proposed in Chap. 6, is required providing optimal transmission control and bandwidth utilization.

Integration of Security analysis with QoS design needs to meet both security and satisfy certain QoS requirements simultaneously from long view. Different from most of the existing works which deal with WSNs strategy to achieve QoS, on Fig. 2.1, we extend QoS support to the model by introducing a number of active security requirements distributed in a gradient fashion based on their logical connection to the QoS requirements.

2.4 Reliability, Availability and Serviceability (RAS)

For availability and serviceability, remote testing and diagnostics is needed to pinpoint and repair (or bypass) the failed components that might be physically unreachable. Severe limitations in the cost and the transmitted energy within WSNs negatively impact the reliability of the nodes and the integrity of transmitted data. The application, itself, will greatly influence how system resources (namely, energy and bandwidth) must be allocated between communication and computation requirements to achieve requisite system performance. Furthermore, although performance of wireless communication systems and communication networks is well understood due to decades of research, the present body of knowledge regarding the performance of WSNs is limited.

However, we examine WSNs nodes and propose the necessary QoS required for increasing both the availability and serviceability of the system, also as a way of intensify security within WSNs. Our approach is service oriented and was particularly motivated by recent proposals of defining QoS for WSNs in previously works.

QoS control is required for the assumption in sense that the number of sensors deployed exceeds the minimum needed to provide the requisite service. It presents two new techniques to maintain QoS under a variety of network constraints [2, 3]. The papers propose a new, extremely low-energy control strategy based on individual feedback in a random access communication system.

With the similar approach on QoS we present a primary application of this phenomenon, to explore and understand a security in WSNs as far as reliability, availability and serviceability are concerned as the indications for DoS detections. In particular, our work is applicable to sensor networks that are deployed in remote, hostile environments e.g., space applications and so on. Such networks are constrained by (1) high die-off rates of nodes and (2) inability to be replenished. The performance of the proposed approach is demonstrated throughout using numerical examples. Reliability of a system is defined as the probability of system survival in a period of time. Since it depends mainly on the operating conditions and operating time, the metrics of Mean Time Between Failure (MTBF) is used. For time period of duration t, MTBF is related to the reliability as follows [17]:

$$Reliability = 1 - \frac{t}{Mean\_time\_between\_failure} \tag{2.1}$$

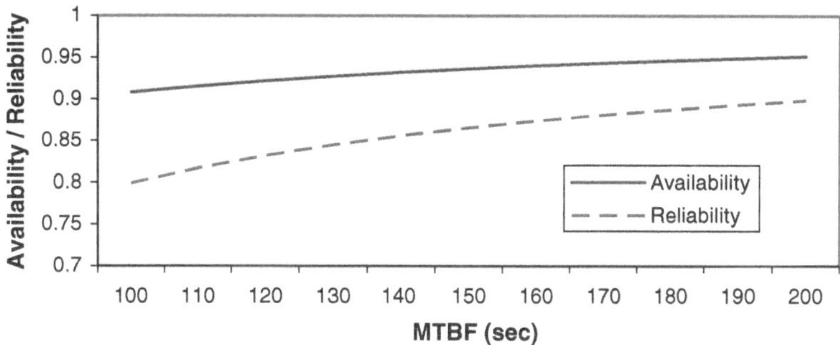

Fig. 2.2 Dependence availability and reliability on MTBF

Availability of a system is closely related to the reliability, since it is defined as the probability that the system is operating correctly at a given time. Dependence availability and reliability on MTBF presented on Fig. 2.2. Calculating availability is related to MTBF and Mean Time To Repair (MTTR) by the following relation [17]:

$$Availability = \frac{Mean\_time\_between\_failure}{Mean\_time\_between\_failure + Mean\_time\_to\_repair} \tag{2.2}$$

Considering availability of each node in isolation, from Eq. (2.2), the MTTR should be minimized, while MTBF should be maximized. While MTBF is given by manufacturing practices and components used, the value of MTTR can be controlled by both individual node and network design.

$$M\,\% = \frac{m \times 100\,\%}{n}$$

where m is a number of failed nodes within WSN, n is number of nodes within WSN and $M\,\%$ is possible percentage of failed nodes within given WSN.

Serviceability of a system is defined as the probability that a failed system will restore to the correct operation. Serviceability is closely related to the repair rate and the MTTR [17].

$$Serviceability = 1 - \exp \times \left(-\frac{t}{Mean\_time\_to\_repair}\right) \tag{2.3}$$

A fundamental service in sensor networks is the determination of time and location of events in the real world. This task is complicated by various challenging characteristics of sensor networks, such as their large scale, high network dynamics, restricted resources, and restricted energy. We use Hawk sensor nodes for determination time of data transmitting in fulfilling the QoS under these constraints. We illustrate the practical feasibility to our approaches by concrete application of real sensor nodes (Hawk Sensor Nodes) to our experiments section.

In any system one must consider the reliability of its components when ascertaining overall system performance. Thus our question was whether the proposed strategy performed adequately for various levels of sensor reliability. Equation (2.1) does not include any information regarding expected sensor life and thus assumes static network resources, which is clearly not the case in WSNs. For example, sensors may fail at regular intervals due to low reliability, due to cost driven design choices, environmentally caused effects (especially in harsh environments), loss of energy, etc.

2.5 Calculating Availability and Probability Within WSN

What we discuss in this section is about achieving two primary factors of dependability in WSNs applications, namely availability and reliability. In the classical definition, a system is highly available if the fraction of its downtime is very small, either because failures are rare, or because it can restart very quickly after a failure [18]. If the application does not require all this redundant information, it would be desirable to conserve energy in some sensors by allowing them to sleep, thereby lengthening the lifetime of the network. For example, as sensors use up their limited energy, the application would like to use different sets of sensors to provide the required QoS (in this case, minimum sensor coverage area). This requires that the application manage the sensors over time. Such management can be as simple as turning sensors on and off, or as complex as selecting the routes for data to take from each sensor to the collection point in a multi-hop network. Furthermore, the needs of the surveillance application may change as a result of previously received data. For example, if the application determines that an intrusion has occurred, the application may assume a new state and require more sensors to send data to more accurately classify the intrusion.

The availability of several implementations is derived from Eq. (2.2) above for MTBF and MTTR. Due to the power issue and the unpredictable wireless network characteristics, it is possible that applications running on the sensor nodes might fail. Thus, techniques to improve the availability of sensor nodes are necessary. Estimated MTBF in our sensor nodes is based on the individually calculated failure rates for each component and the circuit board. Next, for the redundant system versions, if the failure rates (λ) of each redundant element are the same, then the MTBF of the redundant system with n parallel independent elements (i) [19] are taken as:

$$Mean\_time\_between\_failure = \sum_{i=1}^{n} \frac{1}{i\lambda} \tag{2.4}$$

The MTTR can be estimated by the sum of two values, referred to as Mean Time to Detect (MTTD) the failures and the Time to Repair (TTR) ($MTTR = MTTD + TTR$). Notice that this part might be severely affected by the network connections.

Considering the technique [17], where the consumer starts the reparation mechanism by activating the local functional test. Once it completes, the test result is sent back to the consumer for analysis. If a failure occurs, the consumer will send the repair message to the sensor node and initialize the backup component. Acknowledgement is sent back to the consumer once the reparation is completed. If the message latency from the consumer to the target node is *d* seconds and the test time is *c* seconds, then we calculate MTTR as Eq. (2.5):

$$mean\_time\_to\_repair \sim 4d + c \tag{2.5}$$

For the sensor node without the Test Interface Module [17], consumer sends the measured data request command to the suspected sensor node. In order to check the data integrity, same request command will also send to at least two other nearby sensor nodes. The consumer compares the three collected streams of data and pinpoints the failed node. Once the failure is confirmed, consumer will notify the surrounding sensor node to take over the applications of the failed node. Once the failure is confirmed, consumer will notify the surrounding sensor node to take over the applications of the failed node. Again if the message latency from the consumer to the target node is *d* seconds, then MTTR is:

$$Mean\_time\_to\_repair \sim 8d \tag{2.6}$$

To estimate realistic MTTR numbers, we use study [20], where for WSNs Thermostat application with 64 sensor nodes is simulated. Due to the power and protocol requirements, the average latency of related messages is 1522 s. By applying this to our MTTR estimations, the test time *c* is much smaller and can be neglected.

Reliability of a system is defined as the probability of system survival (Fig. 2.6) in a period of time. Therefore, using Poisson probability [21] implemented for WSNs we have as well estimate probability of "failed" situation for whole WSN in given time interval, e.g. for one day (24 h) to demonstrate the reliability of our presented approach.

$$probability(r) = \frac{m^r \times e^{-m}}{r!} \tag{2.7}$$

where *Probability(r)* is a probability of failure system working with "*r*" failed nodes within WSN for given time interval, $r \geq 0$, *m* is a average number of failed nodes within WSN and e = 2.718…

For example, in average there are 3 failed nodes in WSN for 24 h. Then we calculate Probabilities of failure system working as:

$$Probability(\text{"}r\text{"}fails\_for\_24\_hours) = \frac{3^r \times e^{-3}}{r!}$$

$$Probability(0\_fails\_for\_24\_hours) = P(0) = \frac{3^0 \times e^{-3}}{0!} = 0.0498$$

$$Probability(1\_fails\_for\_24\_hours) = P(1) = \frac{3^1 \times e^{-3}}{1!} = 0.1494$$

$$Probability(4\_fails\_for\_24\_hours) = P(4) = \frac{3^4 \times e^{-3}}{4!} = 0.1680$$

From this example, we can see that with progressive increase of fail nodes guantity of a WSN, the risk of unstable work also increases.

2.6 Proposed Security Models

We have proposed thought model, in which things interact once they reach detailed proximity to each other. Here we clarify our assumptions about the flow the QoS uses to map Security. Specifically, the flow model needs to take care of two issues: (1) how to model the relationship between Security instances and QoS; (2) what are the quality factors and sub factors keep track to source QoS. There are of course many options for how to do this; but as far as our approach is concerned in the following part we discuss two reasonable flow ways in Figs. 2.3 and 2.4 for more clarity.

Our model is concern with properties that, such applications must have included availability, reliability, safety and security. The notion of dependability captures these concerns within a single conceptual framework, making it possible to approach the different requirements of a critical system in a unified way as can be seen in Fig. 2.1. We assume that application performance can be described by the quality factor of different variables of interest to the application, where the QoS of the different variables depends on which sensors provide data to the application. For example, in the personal health monitor, variables such as blood pressure, respiratory rate, and heart rate may be determined based on measurements obtained from any of several sensors. Each sensor has a certain QoS in characterizing each of the application's variables. For example, a blood pressure sensor directly measures

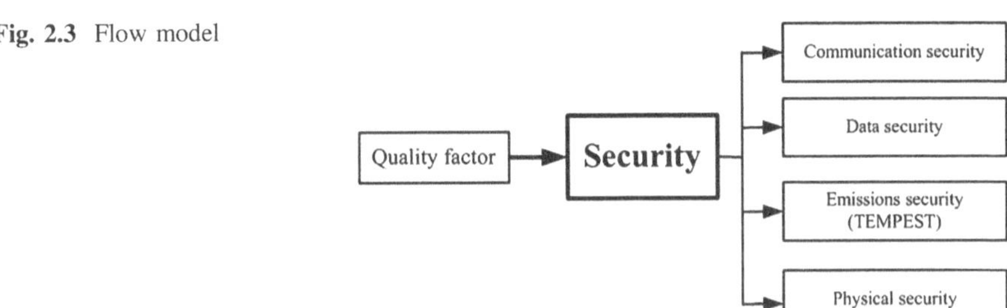

Fig. 2.3 Flow model

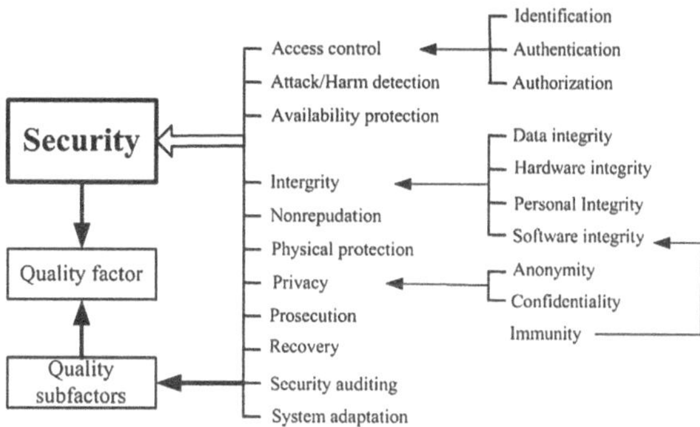

Fig. 2.4 Amalgamation of QoS in WSN security coverage flow model

blood pressure, so it provides a quality of 1.0 (Quality is mapped to a specific reliability in determining the variable from the sensor's data, with 1.0 corresponding to 100 % reliability) in determining this variable. In addition, the blood pressure sensor can indirectly measure other variables such as heart rate, so it provides some quality, although less than 1.0, in determining these variables (data security, communication security, physical security). The quality of the heart rate measurement would be improved through high-level fusion of the blood pressure measurements with data from additional sensors such as a blood flow sensor.

This can be modeled as the application changing state based on Quality factor. For different security states, different sets of sensors should be activated to provide the greatest benefit to the security application.

Figure 2.4 illustrates the important variables to monitor when determining a security condition and indicates the security types that can provide at least some quality to the measurement of these security types. Each line between a security type and a variable is labeled with the quality factor can provide to the measurement of that security.

2.7 Experiment Evaluation

To assess our techniques against traditional security approach in terms of three aspects Reliability, Availability, and Serviceability (RAS) we have implemented various experiments to Pentium III Computer processor and 512 MB memory was used. We measured the processing throughput, i.e., the number of data transmitted events that each phase is able to process per second and time taken to transmit these data within selected sensor nodes, as can be seen in graph presentation in Fig. 2.5.

To simplify the process, we suppose all services share the same QoS as defined formerly. We consider that the element QoS_prediction_input has four properties: Availability, Reliability, Bandwidth, and Request time. Availability measures whether or not the client can connect to the service (i.e., web service, SN service). It takes a value of 0 (can not connect) or 1 (be able to connect). Reliability refers to

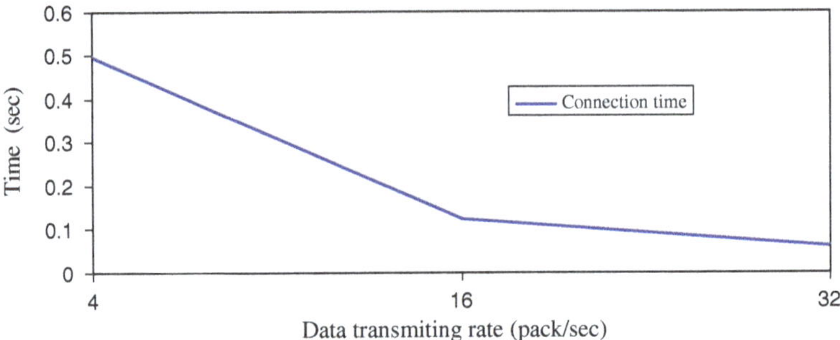

Fig. 2.5 Connection time for 4/16/32 pack/s

whether the operation the client wishes to perform can be performed. It takes a value of 0 (unable to perform the operation) or 1 otherwise. If a service is not reachable, the Reliability is assumed to be 0 for that interaction. Bandwidth is used to measure the network condition, and Request_time is the moment a user requests a particular service. As to the QoS output parameters, we mainly consider two QoS criterions: connection time and transmition_time. The former measures the expected delay between the moments a request is sent and the moment the result is brought back; the latter reflects whether the transmition process has performed properly or not, where 1 means transmition was well performed and 0 means otherwise. The Service's QoS prediction results are given as below however; this methodology requires increased computational complexity.

Notice that the performance of the communication channel is not taken into considerations in the above calculations for single node availability. With channels used for WSNs, packets losses are common. They increase the message latency and can ultimately affect the MTTR. We analyzed further the influence of the network to the availability. We plot the node availability versus average latency, which lumps together the characteristics of the channel, the number of retransmission retries on the failure, as well as the node-dependent features such as retransmission timeouts in Fig. 2.6.

In Fig. 2.6, we examine WSNs nodes to transmit the data in evaluating (RAS). Two sensor nodes with 32 size byte were used for estimating connection time with different transmitting rate. With 0.0625 t/s we were able to connect 32 packets. To ask one sensor node to transmit the data we need 2 data packets (one for asking, another one for receiving the answer). To estimate Time to connection we have to transmit only two packets. *Number of packets = file size/packet size. Time = number of packets/data transmitting rate.* This can be used to propose the necessary infrastructure required for increasing both the availability and serviceability of the system, in spite of the absence of a reliable transport layer. Hence this can be used to analyze and detect delay, delivery, performance or energy consumptions caused by different attacks as elaborated in Fig. 2.1.

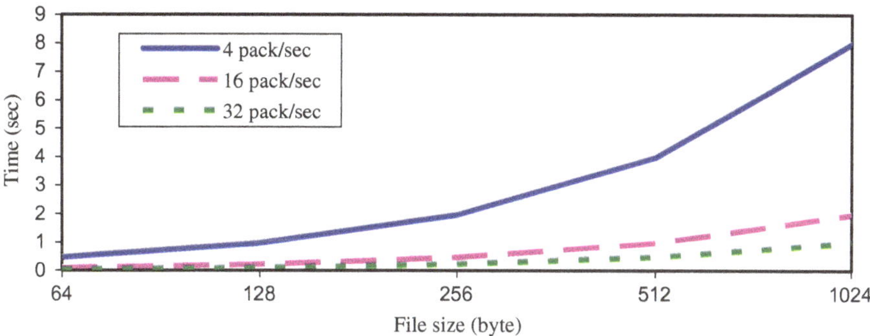

Fig. 2.6 Transmitting time in different number of packets to access (RAS)

2.8 Summary

In this chapter, security of WSN is considered through QoS. Using QoS components, we evaluated models and system-level test using sensor nodes.

One primordial issue is to satisfy application QoS requirements while providing a high-level abstraction that addresses WSN security. Notice that although we consider primarily testing in the lab, the proposed solutions can easily be applied to testing in factory with large size of Sensor network applications. With the proposed approach, such tests can be easily parallelized by applying wireless broadcast to many nodes at once. As a result, the proposed approach can be used in variety of testing scenarios.

Security issues in a health monitoring system utilizing Wireless Sensors in a WSN have been discussed precisely with data integrity in security aspects. A secure model is proposed with flow of security classes providing different levels of security using QoS. However our finding found that effects of security metrics place a lot of burden on the QoS of the overall system thus decreasing performance.

References

1. Chen, D., Varshney, P.K.: QoS support in wireless sensor networks: A survey. In: Proceedings of International Conference on Wireless Networks, vol. 13244, pp. 227–233 (2004)
2. Perillo, M., Heinzelman, W.: Providing application QoS through intelligent sensor management. In: 1st Sensor Network Protocols and Applications Workshop, pp. 93–101 (2003)
3. Iyer, R., Kleinrock, L.: QoS control for sensor networks. In: IEEE International Conference on Communications, vol. 1, pp. 517–521 (2003)
4. Pottie, G., Kaiser, W.: Wireless integrated network sensors. Commun. ACM **43**(5), 51–58 (2000)
5. Estrin, D., Girod, L., Pottie, G., Srivastava, M.: Instrumenting the world with wireless sensor networks. In: Proceedings of International Conference on Acoustics, Speech and Signal Processing, vol. 4, pp. 2033–2036 (2001)

6. MIT Technology Review: 10 emerging technologies that will change the world. MIT Technology Review, Cambridge (2003)
7. Rey, R.F.: Engineering and Operations in the Bell System. Bell Labs, New Jersy (1977)
8. Perrig, A., Stankovic, J., Wagner, D.: Security in wireless sensor networks. Commun. ACM **47**(6), 53–57 (2004)
9. Crawley, E., Nair, R., Rajagopalan, B., Sandick, H.: A framework for QoS-based routing in the Internet. Internet informational RFC 2386, 37. http://wUHHUww.ietf.org/rfc/rfc2386.txt (1998)
10. Chen, S.: Routing support for providing guaranteed end-to-end quality-of-service. Ph.D. thesis, UIUC, 207. http://cairo.cs.uiuc.edu/publications/papers/SCthesis.ps (1999)
11. Min, R., Bhardwaj, M., Cho, S.H., Shih, E., Sinha, A.: Low power wireless sensor networks. In: Proceedings of International Conference on VLSI Design, pp. 205–210 (2001)
12. Stallings, W.: Network security essentials. Applications and standards. Prentice Hall, Upper Saddle River (2000)
13. Karlof, C., Wagner, D.: Secure routing in wireless sensor networks: Attacks and countermeasures. Elsevier's Ad hoc Netw. J. Spec. Issue Sens. Netw. Appl. Protoc. **1**(2–3), 293–315 (2003)
14. Kargl, F., Schlott, S., Klenk, A., Geiss, A.: Michael weber. Securing ad hoc routing protocols. In: IEEE EUROMICRO, pp. 514–519 (2004)
15. Undercoffer, J., Avancha, S., Joshi, A., Pinkston, J.: Security for sensor networks, wireless sensor networks, pp. 253–275 (2004)
16. http://www.csee.umbc.edu/cadip/2002Symposium/sensorids.pdf
17. Chiang, M.W., Zilic, Z., Radecka, K., Chenard, J.S.: Architectures of increased availability wireless sensor network nodes. In: ITC International Test Conference, vol. 43(2), pp. 1232–124 (2004)
18. Knight, J.C.: An introduction to computing system dependability. In: Proceedings of the 26th International Conference on Software Engineering. IEEE Computer Society, pp. 730–731 (2004)
19. Callaway, E.H.: Wireless sensor networks architectures and protocols. Auerbach Publications, UK (2004)
20. Headquarters, Department of the Army, TM-5-698-1: Reliability/Availability of Electrical & Mechanical Systems for Command, Control, Communications, Computer, Intelligence, Surveillance, and Reconnaissance Facilities (2003)
21. Eddousm, M., Stansfield, R.: Methods of decision making. Unity press, Leicester, pp. 50–52 (1997)

Chapter 3
Mathematical Model for Wireless Sensor Nodes Security

3.1 Introduction

WSNs have recently emerged as an important means to study and interact with physical world. In previous era, castles were surrounded by moats (deep trenches, filled with water, and even alligators) to prevent or discourage intrusion attempts. Today one can replace such barriers with stealthy and wireless sensors. In this chapter, we develop mathematical foundations model using barriers concept to design secure wireless sensors nodes. Security becomes one of the major concerns when there are potential attacks against sensor network nodes. Thus we have designed fundamental security in disk shaped to provide basic security elements that can be implemented in various sensor nodes. The mathematical models introduced are flexible and efficient to be embedded in sensor nodes and can create a suitable nodes components security in hostile environments. We also demonstrate how these nodes can be deployed in wall and belt form to fulfill their tasks.

WSNs can replace such barriers today at the building level and at the estate level, where barriers can be more than a kilometer long [1]. Efforts are currently underway to extend the scalability of WSNs so that they can be used to monitor one of the largest international borders [1]. Intrusion detection and border surveillance constitute a major application category for WSNs. A major goal in these applications is to detect intruders as they cross a border or as they penetrate a protected area.

Existing sensor network security research has mostly focused on adapting security mechanisms to the computational and messaging constraints imposed by tiny sensor devices [2, 3]. From an operational point of view, it is also worth mentioning that sensor nodes might or might not have addressable global identification (ID). This fact affects how protocols and security schemes are designed for WSNs. None of the sensor nodes applications would function correctly if appropriate security measures are not taken. Threat such as a mote-class attacker versus a laptop-class attacker, an insider attacker versus an outsider attacker, Passive versus active attacker can be expected from the absence of excellent security mechanisms.

© Springer International Publishing Switzerland 2016

G.S. Oreku and T. Pazynyuk, *Security in Wireless Sensor Networks*,
Risk Engineering, DOI 10.1007/978-3-319-21269-2_3

The security of WSNs can be classified into two broad categories: (1) operational security, and (2) information security. The operation-related security objective is that a network as a whole should continue to function even when some of its components are attacked (the service availability requirement). The information-related security objectives are that confidential information should never be disclosed, and the integrity and authenticity of information should always be ensured. While it may seem that information security can readily be achieved with cryptography, there are 2 facts that make achieving the above objectives non-trivial in WSNs: (1) sensor nodes operate unattended—they are potentially accessible, both geographically and physically, to any malicious party imaginable; (2) sensor nodes communicate through an open medium. The first fact makes insider attacks possible, when the soft belly of every node is up for grabs—it is easy for an attacker to gain access to the data (including system states and cryptographic material) and programs that power the devices, and even modify the software to run its own algorithms.

Effective security support proposed is a model design that needs an integrated approach. Securing a system requires more than simply adding encryption processors and virus-scanning software, rather, you must implement those security elements in an organized way.

A system's security is only as strong as its weakest link. For example, a smart card's strongest cipher algorithm is worthless if a hacker can disassemble the card and retrieve sensitive data by observing its power consumption [4]. We think of security as a design domain with multiple layers of design abstraction, and a complete system as a co-design of domains (security, networking, and graphics, for example) rather than a co-design of implementations (such as hardware and software). The classic view emphasizes using hardware for performance and software for flexibility. In embedded applications such as mobile phones, PDAs, and sensor network nodes, however, energy efficiency is tantamount important.

We approach these challenges by designing domain-specific mathematical based model disc shaped which can be integrated into flexible sensor nodes based on a reconfigurable interconnect sensor network. This type of model is referred to as barrier, where the sensors form a barrier for the intruders.

3.2 Barrier Security

Most of the existing works focus on barrier full-coverage [5–7] and that too in regular regions rather than in a thin belt region. The proofs and the conditions developed for full-coverage do not readily carry over to the case of barrier coverage in thin belt regions.

Another work related to barrier [8] is addressed the issue of intruder tracking in regular regions such as a square. The focus of this work is the following problem—given a value of l what is the minimum number of sensors needed so that if the nodes are independently and uniformly distributed, the average length of an uncovered path traveled by an intruder that starts at a random (uniformly chosen) location within the field. Although this is an important problem for tracking applications, it does not address the problem sensor nodes design in mathematical model as means creating robust security.

The concept of barrier coverage first appeared [9] in the context of robotic sensors. Simulations were performed in [10] to find the optimal number of sensors to be deployed to achieve barrier coverage. To the best of our knowledge, ours is the first work to address the theoretical foundation for disc sensing design that can be employed as barrier (using critical WSNs conditions) to achieve security in sensor network.

As can be seen from the discussion of some related work above, a lot of interesting works have come close to the problem of barrier coverage, but none have addressed the issue of barrier design to deriving critical security conditions for WSNs, which is a more realistic model for sensor deployed on unsecured environments. Also, no existing work, to the best of our knowledge, has addressed the issue of developing efficient mathematical design for determining whether a given node barrier can stop malicious action on sensor network.

Earlier research on sensor networks has focused on developing extremely optimized protocols at different layers of networking stack, as well as a specialized operating system called TinyOs [11]. However, the majority of these protocols have not been designed with security and privacy in mind resulting in substantial performance degradation if there is a security breach. Security can not be designed as a separate module to be added on top of these protocols. Rather, security has to be integrated in the design of every component of the sensor network.

3.3 Problem Statement and Mathematical Model Design

Security becomes one of the major concerns when there are potential attacks against sensor networks. Many protocols and algorithms (e.g. routing, localizations) will not work in hostile environments without security protection. Security services such as authentication and key management are critical to ensure the normal operations of sensor network in hostile environments.

In this section we present our secure mathematical based model designed to be implemented in sensor nodes in disc shape creating barrier to intruders who might attack sensor nodes. The design can be implemented to sensor nodes and its formalizations are elaborated in details in this section as well. Assume that each sensor node has only secured protocol, or locations. An attacker may capture or compromise one or number of sensor nodes without being noticed if nodes are not secured. If the sensor nodes are compromised, the attacker learns all the secrets stored on them and may launch a variety of malicious actions against the network through these compromised nodes. For example a compromised node may discard all important messages in order to hide some critical events from being noticed or report observations that are significantly different from those observed by non compromised nodes in order to mislead any decision made on these data. The results will be worse if the nodes that provide some critical functions (e.g. data aggregation) are compromised.

The key issue here is to develop a secure mathematical model which can be used to cover every point of security when it comes to WSNs security, not only protocols carrying the information within sensor nodes but as well as node itself and make sure WSNs as a whole are well protected and are resilient to node compromise attack in the sense that no even one or number of sensor nodes can be compromised and sensor network function correctly. Our model can be a proper application to the work well developed [12] with high probability guarantees the detection of intruders as they cross a barrier of stealthy sensors, a sensor network providing strong barrier coverage with high probability (at the expense of more sensors) guarantees the detection of all intruders crossing a barrier of sensors, even when the sensors are not stealthy.

3.4 Formalization

Figure 3.1 represents the essence of formalization of passive countermeasure against attacks, it shows how operation might take place and how the defense mechanism can be implemented (the number of barriers which can be implemented are from *1*st to *K*th). Data, commands and the messages can be presented in the capacity of "Nodes information".

For clarity we present our formalization in elementary security model in Fig. 3.2 which showed that there is locked secure contour around protected information.

Defense stability depends on defense's properties. The principal role is the defense ability to resist overcoming attempts sent by attacker. Here we present two ways to estimate sufficient defense's stability:

- If cost of measures to overcome the defense is more than cost of secured information, then we count defense's stability is sufficient;
- If the timetable to overcome the defense is more than information lifetime, then defense's stability is sufficient.

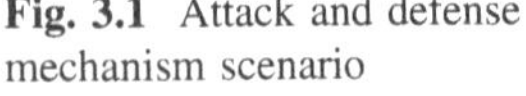
Fig. 3.1 Attack and defense mechanism scenario

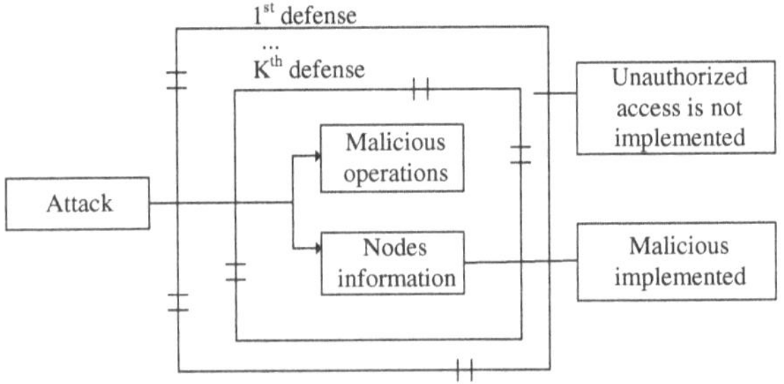

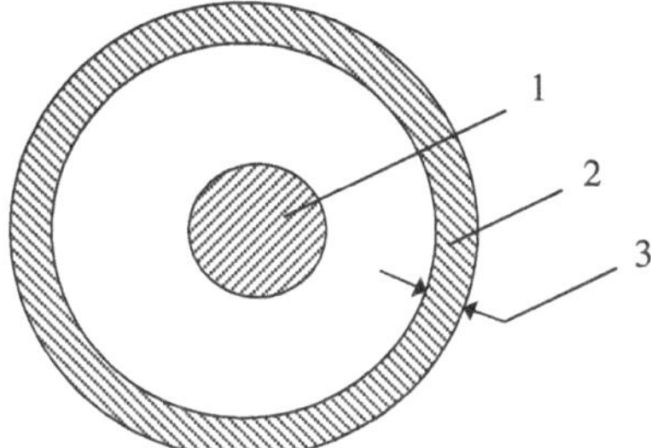

Fig. 3.2 Elementary security model. *1* Secured information; *2* barrier defense mechanism; *3* barrier defense stability

Indication

P_{stable} Probability of defense stability (probability of a barrier's irresistibility);
t_{lt} Information lifetime;
t_{ex} Expected time to overcome the defence by attacker;
P_{break} Probability to break the defense.

However we formulate our second case (timetable to overcome the defense is more than information lifetime) as follow:

- $P_{stable} = 1$ if $t_{lt} < t_{ex}$ and $P_{break} = 0$.
  $P_{break} = 0$ indicates that there is locked secure contour around protected information (system is stable).
- If $t_{lt} > t_{ex}$ and $P_{break} = 0$, then $P_{stable} = (1 - P_{br\_\min})$,
  where $P_{br\_\min}$—probability of overcoming the defence by the attacker for time less than t_{lt}.

In actual conditions there are $t_{lt} > t_{ex}$ and $P_{break} > 0$, therefore we can estimate our security strength by Eq. (3.1)

$$P_{stable} = (1 - P_{br\_\min}) \times (1 - P_{break1}) \times \cdots \times (1 - P_{breakK}) \tag{3.1}$$

where
$P_{br\_\min} = 0$, if $t_{lt} < t_{ex}$;
$P_{br\_\min} > 0$, if $t_{lt} \geq t_{ex}$;
K—number of ways to break the defence, i.e. for each barrier it can be several ways to be overcomed.

The choice and definition of P_{break}, firstly, can be made by expertise way on basis of previous experience. P_{break} must take on value from 0 to 1, otherwise with $P_{break} = 1$ the effect of security is lost.

When the secured information is refresh periodically, i.e. with $t_{lt} > t_{ex}$, permanent defense is used which can discovers and blocks the access of attacker to the secured information.

Our proposed automated defense's principle is based on following: Periodically, control module monitors all the sensors to find the overcomings. The condition for defence stability with disclosure and blocking of unauthorized access can be presented as follow (3.2):

$$\frac{T_{inquiry} + t_{response} + t_{ld} + t_{block}}{t_{ex}} < 1 \quad (3.2)$$

or

$$\frac{T_{total}}{t_{ex}} < 1 \quad (3.3)$$

where

$T_{inquiry}$	Sensor's inquiring period;
$t_{response}$	Disturbed signalization response time;
t_{ld}	Location disclosure time;
t_{block}	Access blocking time;
$T_{total} = T_{inquiry} + t_{response} + t_{ld} + t_{block}$	Disclosure and blocking time of unauthorized access.

We present an unauthorized attacker's actions in form of temporal graph presented on Fig. 3.3. Here we assume that time interval T is T = (¼ of the total time), i.e. $T = \frac{1}{4} T_{total}$. Our results show that the response time ($t_{response}$) is excellent variable as it can counter unauthorized access as well as attacker within ¼ of total time before even their disclosure or blockage. However, the approach is efficient as adversaries can be revealed or blocked before they destruct the system within given total time.

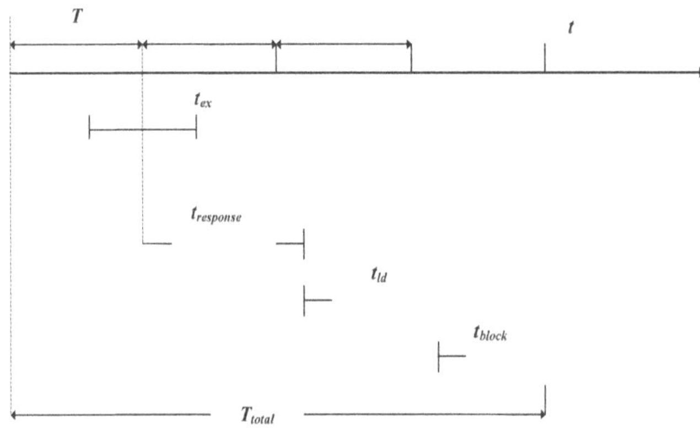

Fig. 3.3 Temporal diagram of unauthorized access control

From Fig. 3.3 we can see, that adversary could not be disclosured into the two cases:

- $t_{ex} < T$;

  Probability of attack to succeed ($P_{br\_\min}$):

$$P_{br\_\min} = \frac{T - t_{ex}}{T} = 1 - \frac{t_{ex}}{T} \quad (3.4)$$

  Probability of discovering malicious action (P_{md}):

$$P_{md} = 1 - P_{br\_\min} \quad (3.5)$$

  or

$$P_{md} = \frac{t_{ex}}{T} \quad (3.6)$$

- $T < t_{ex} < T_{total}$;

  Probability of attack to succeed (P_{br_min}):

$$P_{br\_\min} = \frac{T_{total} - t_{ex}}{T_{total}} = 1 - \frac{t_{ex}}{T_{total}} \quad (3.7)$$

  Probability of discovering and blocking malicious action (P_{mdb}):

$$P_{mdb} = 1 - P_{br\_\min} \quad (3.8)$$

  or

$$P_{mdb} = \frac{t_{ex}}{T_{total}} \quad (3.9)$$

3.5 Ensuring Passive Resistance to Threat on Sensor Nodes

Concept of defending sensor nodes using in practice, protective contour or barrier designed consists of coupled of connected barriers with different strength and to the best of our knowledge have never been studied before. The defense mechanism also includes several barrier's strength designed in layer form in disc shape as can be seen in Figs. 3.4 and 3.5. We consider our sensor being scattered randomly in the field and will form a sensor network after deployment in an ad hoc manner to fulfill certain tasks and eventually they will have multiple links. Each individual sensor

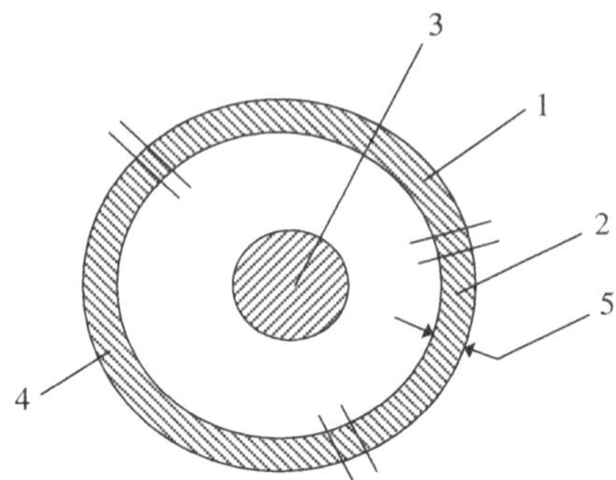

Fig. 3.4 Model of multilink security. *1* Barrier 1; *2* barrier 2; *3* secured information; *4* barrier 3; *5* barrier strength

node then monitors conditions and activities in its local surrounding and reports its observations to central server by communicating with its neighbors. Obviously, the design of sensor nodes requires wireless networking techniques, especially wireless ad hoc networking techniques. As most traditional wireless networking protocols and algorithms are not suitable for sensor network [2, 13–15] we propose the use of mathematical model of barrier's defense to be integrated in sensor nodes for security purposes. The protective contour consisted of coupled of connected barriers (barrier 1, 2 and 3) with different strength can be called as "Multilink model" and presented as follow (Fig. 3.4).

Since we have multiple links we calculate probability of our multilink security (P_{stable}) as:

$$\begin{aligned} P_{stable} &= P_{stable1} \times \cdots \times P_{stable_n} \\ &= (1 - P_{break_1}) \times \cdots \times (1 - P_{break_n}) \end{aligned} \tag{3.10}$$

where P_{stable_n} is strength of nth barrier and P_{break_n} is probability of nth barrier breaking.

We also consider if stability of weakest part (barrier) is satisfy to qualifying standards of security requirements in general, then we will have the redundant strength of another parts of contour. Therefore, the use of equal-stable barriers will be economically reasonable in multilink security contour.

With the higher requirements of security, we propose to use multilevel security model as it presented in Fig. 3.5. Here we use several contours (levels) to achieve greater security of secured information. Figure 3.5 presented three security levels—level 1, 2 and 3. Also, each level still can have one or more connected barriers as it was presented at "Multilink security model" (Fig. 3.4).

The total strength of security contours (P_{Σ}) to our proposed multilevel security model can be presented mathematically as:

$$P_{\Sigma} = 1 - \prod_{i=1}^{m} (1 - P_{stable_i}) \tag{3.11}$$

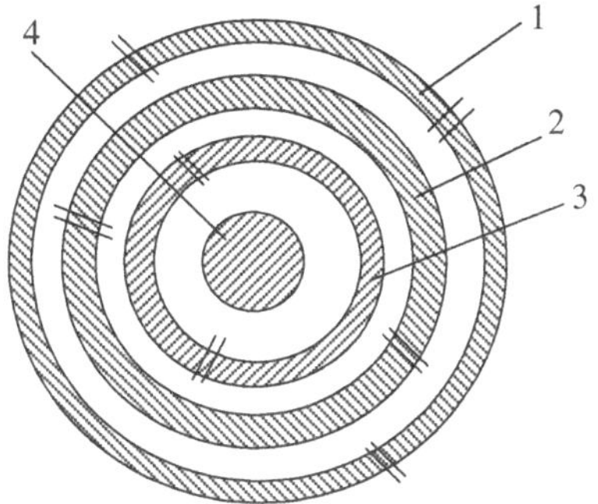

Fig. 3.5 Multilevel security model. *1* 3rd security contour; *2* 2nd security contour; *3* 1st security contour; *4* secured information

where $1 \leq i \leq m$—ordinal (serial) number of barrier; m—number of levels ($m = 3$ in the case of Fig. 3.5); P_{stable_i}—*i*th contour's stability. So, to calculate P_{Σ}, firstly, we should calculate each level strength using Eq. (3.10).

With $P_{stable_i} = 0$, there is no need to calculate *i*th contour's strength. With $P_{stable_i} = 1$, others security contours are redundant. This model suits only for security contours, which block the same unauthorized access channel to the same subject of security.

3.6 The Nodes Model Application

In this section we present two ways of implementation our secure model, there are *k*-lines covered area and belt secure region (Figs. 3.6 and 3.7).

Assumption 1 (*Mathematical-based nodes*) We assume a mathematical node based model where each active sensor nodes has a sensing radius of *r*; any object within the nodes of radius *r* centered at an active sensor network is reliably detected by it. The sensing sensor node located at location *u* is denoted by $D_r(u)$.

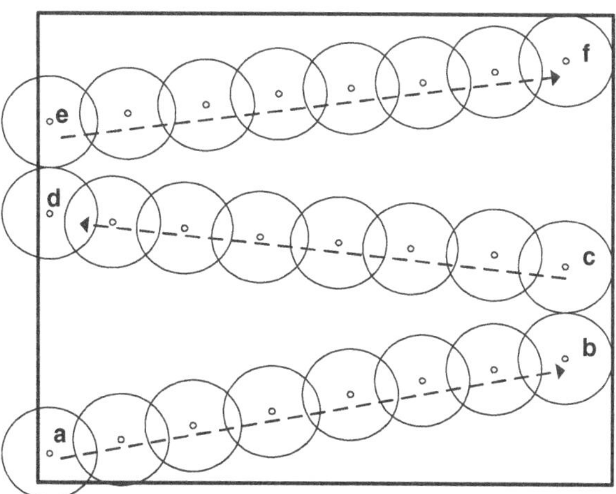

Fig. 3.6 The above region is 3-lines nodes covered area

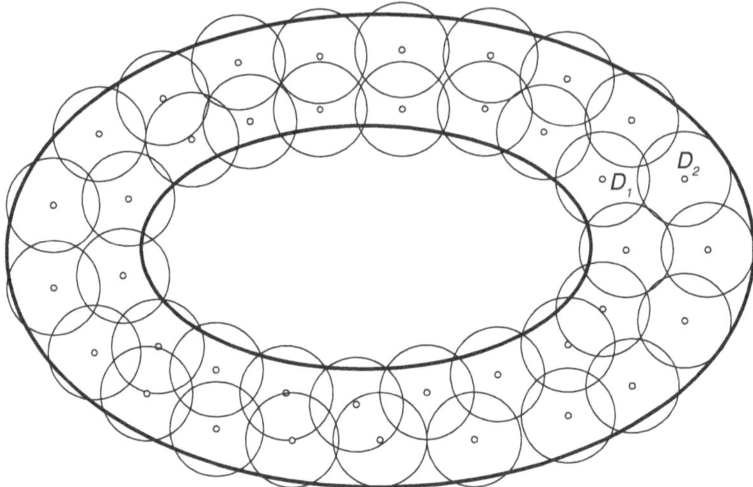

Fig. 3.7 A sensor network deployed over a closed belt region

Definition 1 (*Intruder*) An intruder is any person or object that is subject to detection by the sensor network nodes as it crosses the barrier.

Looking at the sensor deployment in Fig. 3.6, one can easily conclude that the region is 3-barrier covered (from node *a* to node *b*, from node *c* to node *d* and from node *e* to node *f*) since there does not exist any path that crosses the complete width of the region without being detected by at least three sensors (nodes *a*, *d*, *e* from the left border and nodes *b*, *c* and *f* from the right border).

Assumption 2 There is the belt region (B) in consideration (Fig. 3.7). If two sensing node D1 and D2 have overlap, then $(D_1 \cup D_2) \cap B$ is a connected sub-region in B.

In the Fig. 3.7 we show the resulting coverage graph for the sensor networks presented as a belt region, covered by two closed barriers. The area is well secured since there does not exist any path that crosses the complete width of the region without being detected by at least sensors nodes.

3.7 Summary

Detection of intruders breaching the perimeter of a building or an estate, or those crossing an international border is increasingly being seen as an important application for WSNs. We need a theoretical foundation to determine the minimum number of sensors to be deployed so that intruders crossing a barrier of sensors will always be detected. However the sensor nodes deployed should have the security implemented in them as suggested with our findings.

In this chapter, we present the fundamental mathematical model design for sensor nodes that can be used to secure different WSNs topology against intruders. As we are still in the early stage of our findings the concept of barrier technique design in sensor nodes is a relatively new concept, several problems still remain open in this space. One such problem is the implementation of our approach to already existing barrier coverage presented techniques. Another open problem is the impact of our mathematical model in sensor nodes is not yet fully explored. In our future work, we plan to address these and other open problems in the area of sensor nodes design with respect to our presented model.

References

1. Extreme scale wireless sensor networking. Technical Report. http://www.cse.ohio-state.edu/exscal/ (2004)
2. Perrig, A., Szewczyk, R., Wen, V., Culler, D., Tygar, J.D.: SPINS: Security protocols for sensor networks. In: Proceedings of the 7th Annual International Conference on Mobile Computing and Networking, pp. 189–199 (2001)
3. Karlof, C., Sastry, N., Wagner, D.: TinySec: Link-layer encryption for tiny devices. In: Proceedings of the 2nd ACM SenSys Conference (2004)
4. Anderson, R., Kuhn, M.: Tamper resistance—A cautionary note. In: Proceedings of 2nd Usenix Workshop Electronic Commerce, Usenix, pp. 1–11 (1996)
5. Huang, C., Tseng, Y.: The coverage problem in a wireless sensor network. In: ACM International Workshop on Wireless Sensor Networks and Applications (WSNA), pp. 115–121 (2003)
6. Kumar, S., Lai, T.H., Balogh, J.: On k-coverage in a mostly sleeping sensor network. In: International Conference on Mobile Computing and Networking, pp. 144–158 (2004)
7. Zhang, H., Hou, J.: On deriving the upper bound of α-lifetime for large sensor networks. In: International Symposium on Mobile Ad Hoc Networking and Computing, pp. 121–132 (2004)
8. Gui, C., Mohapatra, P.: Power conservation and quality of surveillance in target tracking sensor networks. In: International Conference on Mobile Computing and Networking, pp. 129–143 (2004)
9. Gage, D.W.: Command control for many-robot systems. In: Nineteenth Annual AUVS Technical Symposium, vol. 10(4), pp. 28–34 (1992)
10. Hynes, S.: Multi-agent simulations (mas) for assessing massive sensor coverage and deployment. Technical Report, Master's Thesis, Naval Postgraduate School (2003)
11. Gay, D., Levis, P., Culler, D.: Software design patterns for TinyOs. In: Proceedings of the ACM SIGPLAN/SIGBED Conference on Languages, Compilers, and Tools for Embedded Systems, pp. 40–49 (2005)
12. Kumar, S., Lai, T.H., Arora, A.: Barrier coverage with wireless sensors. In: MobiCom, pp. 284–298 (2005)
13. Niculescu, D., Nath, B.: Ad hoc positioning system (APS). In: INFOCOM 2003. IEEE Twenty-Second Annual Joint Conference of the IEEE Computer and Communications Societies, pp. 1734–1743
14. Intanagonwiwat, C., Govindan, R., Estrin, D.: Directed diffusion: A scalable and robust communication paradigm for sensor networks. In: Proceedings of the Sixth Annual International Conference on Mobile Computing and Networking, pp. 56–67 (2003)
15. Newsome, J., Song, D.: GEM: Graph embedding for routing and data-centric storage in sensor networks without geographical information. In: Proceedings of the First ACM Conference on Embedded Networked Sensor Systems, pp. 76–88 (2003)

Chapter 4
Improved Feistel-Based Ciphers for Wireless Sensor Network Security

4.1 Introduction

WSN are exposed to variety of attacks as other networks. Quality and complexity of attacks are rising day by day. The proposed work aim at showing how complexity of modern attacks is growing accordingly, converging to usher resistant methods also to rise. Limitations in computation and battery power in sensor nodes gives constrain to diversity of security mechanisms. We must apply only suitable mechanism to WSN where by applications of improved "Feistel Scheme" motivated our approach. The modified accelerated-cipher design use data-dependent permutations (DDP), and can be used for fast hardware, firmware, software and WSN encryption systems. The approach presented showed that ciphers using this approach have less intrusion probability against differential cryptanalysis than currently used popular WSN ciphers like DES, Camellia and so on.

The goal of information security is to provide information safety and integrity [1, 2]. Information transferring through WSN needs to be protected from misuse respectively. Modern security methods need to guarantee safety of data transmitting with respect to security needs i.e. Confidentiality, Integrity, and Availability (CIA). Providing information security in WSN is also necessary especially for those security-sensitive applications and is one of the major concerns to our proposal. There are lots of countermeasures methods have been extensively studied to provide WSN communication securities [1, 3–5]. However WSN is still exposed to some kinds of attacks [4–6]. These defenses are ineffective against attacks i.e. from compromised servers due to WSN level constantly increasing, attacks and becoming more and more complicated [2, 5, 6]. Moreover WSN has some restrictions when it comes to its applications like limited power supplies, low bandwidth, small memory sizes and limited energy which make it more vulnerable [7]. And as information become more valuable and costly making intruders to use more complicated methods in attacking WSN, eventually this makes security issue become highly sensitive. Due to the increases of new trend of attacks previously

© Springer International Publishing Switzerland 2016

G.S. Oreku and T. Pazynyuk, *Security in Wireless Sensor Networks*, Risk Engineering, DOI 10.1007/978-3-319-21269-2_4

security methods are unable to combat or resist against modern attacks. We present additional step to create efficient and resources constrained security mechanisms for WSN.

Our study shows that new and more stable security approaches need to be put in place to provide information safety considering the following attributes: availability, confidentiality, integrity, authentication, and non-repudiation. We propose to use a modified accelerated-cipher using permutations (DDP) presented as a cryptographic primitive approach for WSN. This concept of DDP is perspective approach in many information securities today [8–10]. Constituting to the key challenges we follow this approach by using Feistel scheme approach to present our improved cipher block using DDP. By cryptanalysis realization, it's necessary to consider differential and linear properties of individual round transformation crypto primitives of block ciphers. This method allows us to create more stable secure mechanisms against modern types of attacks and also provide high-accelerated security program within small sensor devices. In this presentations we use controlled permutation boxes based method for block ciphers implementation to provide modified stable cipher against modern crypto-attacks such as differential cryptanalysis in WSN. The proposed cipher is free key preprocessing which provides high performance in frequent keys exchange. In our work we show the effectiveness of using DDP in ciphers design for WSN. Experiments presented in the rest of this chapter demonstrate the best results of DDP-based ciphers.

4.2 Attacks Threats

Methods of crypto attacks are very complicated. They combined mathematics, information science and even electronics with non ordinary thinking. WSNS block ciphers design needs to consider stability against analytical crypto-attacks. Past years practices has shown that differential (DCA) [11] and linear cryptanalysis (LCA) [12] where the most powerful analytical crypto analysis methods used. The main content of DCA is analysis of influence propagation degree in plaintext modification at cipher text (propagation properties). Using DCA as one of complex attack with complicated mathematics methods can be one of proof verifying to block ciphers stability.

In block cipher cryptanalysis realization, it's necessary to consider differential and linear properties of individual round transformation crypto primitives of blocks. The cases are complicated to element addition on stable round transformation which sometimes might give the negative results to a given cipher algorithm. Block cipher designers who are trying to use theoretical computing constructions that provided distinctness at block ciphers evaluation to modern cryptanalysis methods should give consideration before putting all these into action [13].

Besides differential cryptanalysis there are many threats against new modern networks. One of the main challenges is the design of these networks and their vulnerability to security attacks which leads to network destruction and poor

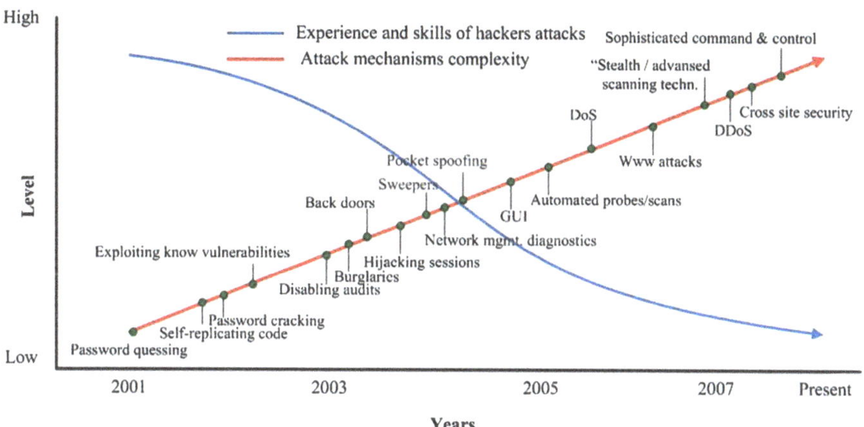

Fig. 4.1 Advancement of complex attacks level, mechanisms and hacker's

performance. Every year the attack complexity increases as can be seen on presented graph in Fig. 4.1.

Figure 4.1 shows attack increasement and complex mechanisms against hacker's skills and experience. Every year not only quantity and complexity of new threats are rapidly increasing but also appearance and momentum. Resistance against them is becoming more and more complicated. Malicious are using more of these security vulnerabilities especially to attack WSN due to the wireless weakness in security.

4.3 The Efficiency of Existing WSN Algorithms

We outline brief draw backs of existing algorithm methods which are being used in many current technologies.

- Widespread algorithms (End to end, single destination communication, IP overlays);
- Probabilistic broadcasts (Discrete effort: does not handle disconnection);
- Scalable Reliable Multicast (Multicast over a wired network, latency-based suppression);
- SPIN (Propagation protocol, does not address maintenance cost) [14];
- Public-key cryptography is too expensive to be usable;
- Fast symmetric-key ciphers must be used sparingly [1].

On designing WSN protocol it's necessary to consider all WSN specific features. For example, communication bandwidth is extremely limited in these networks: each bit transmitted consumes about as much power as executing 800–1000 of operation instructions, and as a consequence, any message expansion caused by security mechanisms comes at significant cost [1, 15].

However we present sets of requirements to WSN protocols [14]. We use these requirements as the highlight in facilitate the design of our new improved cipher.

- Low maintenance overhead (Minimize communication when everyone is up to date);
- Rapid propagation (When new data appears, it should propagate quickly);
- Scalability (Protocol must operate in a wide range of densities, cannot require a priori density information);
- Technical cryptanalysis stability (high-frequency influence at sensors with the purpose of information distortion. These methods allow to get rounds keys value. Last researches showed block ciphers are instable to this kind of attack).

4.4 Techniques of Proposed Method

The presented techniques are based on original Feistel scheme which due to its significant properties can be used in WSN security applications. The modified Feistel scheme design is capable of meeting today's security challenges and generate high-quality results.

4.4.1 *Feistel Scheme*

In all of WSN's blocks ciphers designed by 16-rounds Feistel scheme, data block coding are realizing by two sub blocks using data transformation and *F* function (round encoding function). Like many other symmetric block ciphers DES is also a Feistel Network [11]. The name comes from Horst Feistel who first proposed such a network in early 1970s. In a Feistel network the plaintext is divide into two halves fro the first round of computations which is repeated a number of times (i.e., in a subsequent rounds). Generally the output of the *i*th round is determined from the out of the previously round in the following way (Eqs. 4.1 and 4.2):

$$L_i = R_{i-1} \tag{4.1}$$

$$R_i = L_i \oplus f(R_{i-1}, K_i) \tag{4.2}$$

where *f*() represents the round function, K_i key of *i*th round, L_i and R_i are the left and right parts of data block of *i*th round.

The advantage of Feistel scheme is that block cipher used is very difficulty to breach by proportional of one round key (2^{m}) enumeration [9]. So to determine the requirements for one round cipher transformation during Feistel scheme design is necessary. We briefly indicate below the essential need for designing:

- Increase size of transcriptive block up to 128 bits and more;
- Increase round key size;
- Provide round key elements inseparability within the limits of one algorithm round;
- Using the special methods which prevent mathematical and technical analysis especially addition of some transformations at the beginning of the algorithm and after last round.

Nevertheless before implementing Feistel Schemes to network security we will also like to analyze cons and pros of this approach to network in few words as follow:

Advantages of Feistel approach to networks

- In Feistel scheme we can encode and decode by one operations sequence. Encoding algorithm modification is achieved by queue of round sub keys using modification;
- It minimizes software coding.

Disadvantages of Feistel approach to networks

- In Feistel scheme we have two parts, left and right but only one part of block is used for coding in one round. For example, if block on right side (*R*) is used for the first time in coding the second one on the left part (*L*) is only use for exchanging places, thus not all parts of block are participating in coding process;
- Transformation is very simple because of round function *F* depends only on two parameters (*L* and round key *R*).

For understanding of our presentations we give further destabilizations in this paragraph, given Feistel scheme (Fig. 4.2) one of the standards we elaborate in

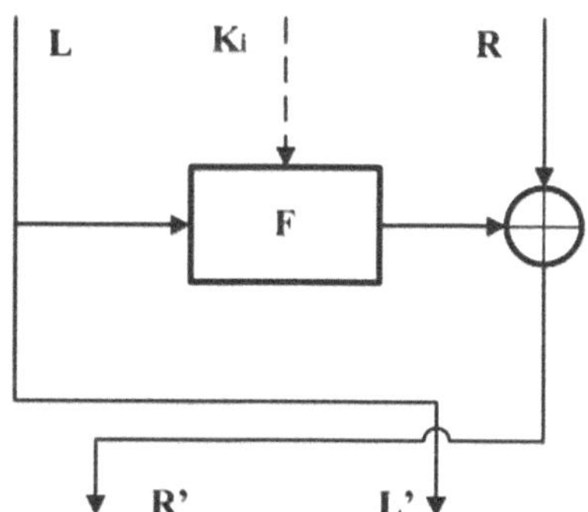

Fig. 4.2 One round modified Feistel scheme

details how Feistel scheme works. Right part R' of transcriptive data $L'||R'$ is a result of group operation $XOR(\oplus)$ where F_{K_i} is a ith round function, i is a round number and K_i is a round key: $R' = R \oplus F_{K_i}(L)$. For advance readings and details about Feistel scheme one should see [16] as well.

Feistel scheme appeared much earlier than modern crypto-attacks as the original cipher using block structure. Its modified version is applied further to limited resources devices as well as embedded devices. From the original standard version it is seen that unmodified version does not meet new security requirements paradigm. The latest record in cracking DES (as of September 1999), set by the Electronic Frontier Foundation's "Deep Crack" is 22 h and 15 min [17]. It involved about 100,000 PCs on the Internet. It was performed as a "know cipher text attack" based on a challenge from RSA Laboratories. The task was to find a 56-bit DES key for a given plaintext and a given cipher text. More so this is well demonstrated in Fig. 4.1 presented above, that no matter many securities being installed in different places but every year attacks trends are strongly increasing in many computer applications. Taking Feistel approach as a key to our methodology we present our modified version to meet the new attack challenges in section "An improved Feistel scheme for block data transformation".

4.4.2 Theoretical Approach of CPB

In our work we propose using controlled permutation boxes for implementation of Feistel scheme design for WSN security. Data depend permutations (DDP) can be performed with so called controlled permutation boxes (CPB) which are fast if implemented in cheap hardware. CPB is one parts of comprehensively upcoming commencement of controlled operation in security applications [18].

The main content of this concept is to created substitution and permutation elements of block ciphers. They provide high-accelerated program-realization nonlinear transformations with small volume of modifications. These transformations are realized by the whole large size of data block at once (32 and more bits) and managed by transcriptive data and algorithm's keys dynamically. CPB mechanisms and its implementation in block ciphers methods provide high stability of such algorithms to modern crypto-attacks such as differential cryptanalysis [9].

WSNs use the block-algorithms data encryption for data transfer. Quality of these algorithms depends on indexes of binary information "dispersion" and "interfusion" which provide interchange of substitution and permutation transformations [11]. In the modern block ciphers these transformations are used by applying two types of crypto primitives:

- Special nonlinear S-box given at the table view. S-boxes provide degree of each block nonlinearity and degree of errors propagation. But small size of S-boxes also gives inconveniency for encoding data block to achieve high indexes on the

following parameters: nonlinearity degree, errors propagation degree and correlation insusceptibility level [11].
- Standard arithmetic or algebraic operations realized with computer commands. Arithmetic operations are effective in software implementation and not complicated in hardware implementation. They have high correlation insusceptibility for all encoding blocks but low degree of nonlinearity and errors propagation.

Modern approach does not give guarantee to maximum security in using Feistel scheme as they have some disadvantages. Attempting to solve this problem we employ controlled operations to make important adaptation of controlled permutation boxes. Controlled operations are described as more simple operations multitudes that are being selected depending on some controlling code value. Controlled permutation boxes (CPB) are alternative to traditional S-boxes and common mathematic operations that generally used at block cipher synthesis [9]. Thus availability of special crypto primitive creation is becoming obviously. These crypto primitives combine and optimize advantages of block ciphers substitution transformations.

4.4.3 An Improved Feistel Scheme for Block Data Transformation

In this sections we consider one round of Feistel scheme with CPB (Fig. 2.3a). In an improved scheme right part R' of encrypted data block can be calculated as: $R' = G_U^{-1}(G_V(R) \oplus F_{K_i}(L))$, where G_V and G_U^{-1} are mutually inverse transformations and depend on control vectors V and U, i.e. G_V, XOR and G_U^{-1} transformations are implementing in series. Generally, control vectors V and U are values of some procedure E from two variables (Fig. 4.3b): data block L and round key K_V(or K_U), i.e. $V = E(L, K_V)$ and $U = E(L, K_U)$. Highest possible unity number $\lambda_{\max}(||A'||)$ for given scheme is also $n^2/2 + n/2$, but here independence between categories of output block R' is achieved greatly easer. Two mutually inverse transformations G_V and G_U^{-1} are provided possibility of using one scheme for direct and inverse transformations, but keys order using is more complicated.

Figure 4.3 shows the main concept of implementing CPB in Feistel scheme. In our work some ciphers based on CPB have been mentioned as well for later comparison in experimental performance. The more detailed information about Cobra-F64a, Cobra-F64b and Spectr-H64 with Feistel characteristics can be found in [18–21].

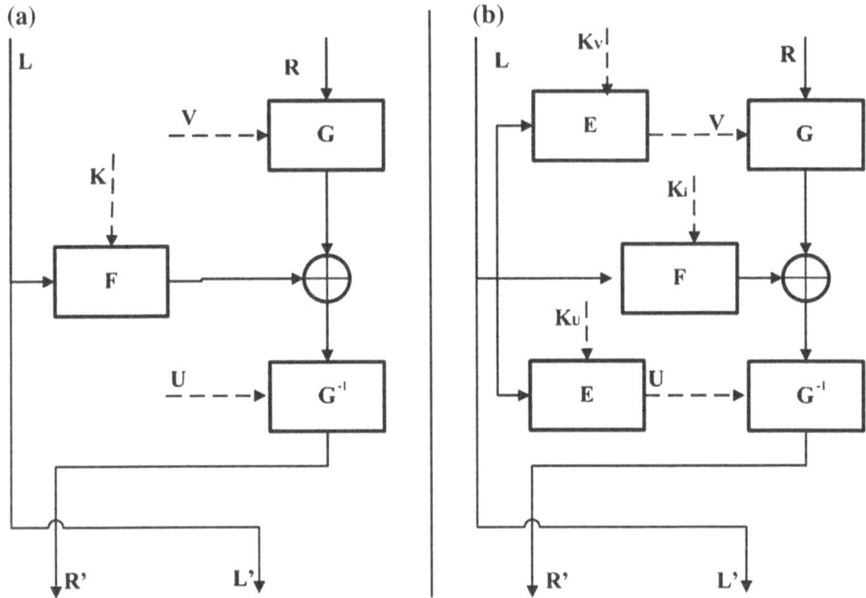

Fig. 4.3 One round scheme of basic (**a**) and detailed (**b**) improved Feistel

4.5 Comparison of CPB-Feistel Scheme Based Ciphers Versus Ciphers with No CPB

Improved Feistel-scheme with different variations of DDP can be implemented in some encoded WSN algorithms, especially for effectiveness of hardware implementation and nature of its block ciphers which basically fits packet structure that can be transmitted within WSN. In case of embedded devices implementation, effectiveness can be achieved from SPECTR and Cobra-ciphers which are CPB based as well. They provide performance of about 20 Mbit/s for microcontroller working at 30 MHz [9]. We run experiment and compare our improved Feistel scheme performance and its stability for data security in different versions of Feistel-based ciphers, i.e. Cobra-F64a, Cobra-F64b, Cobra-S128, Spectr-H64 [9], Camellia [22] and DES [23] against differential cryptanalysis and we present our results in Table 4.1.

Table 4.1 and Fig. 4.4 show the results of differential cryptanalysis security estimation of ciphers, CPB-based block-ciphers, Cobra-S128, Cobra-F64a, SPECTR-H64 and Cobra-F64b. Camellia and DES are examples of block—ciphers based on traditional Feistel scheme. We can see that DDP-based ciphers have more security capability due to less probability of breaking against differential cryptanalysis (in Fig. 4.4 DES and Camillia show a higher probability to be broken than another ciphers). Obtained results show that all considered ciphers are secure against differential crypto—attacks and DDP-based ciphers perform better.

Table 4.1 Differential cryptanalysis security estimation

Cipher	Max number of rounds	Number of round	Probability of attack success
Cobra-S128	12	2	2^{-32}
Cobra-F64a	16	3	2^{-21}
SPECTR-H64	12	2	1.1×2^{-13}
Cobra-F64b	20	2	2^{-12}
Camellia	24	3	2^{-12}
DES	16	2	2^{-7}

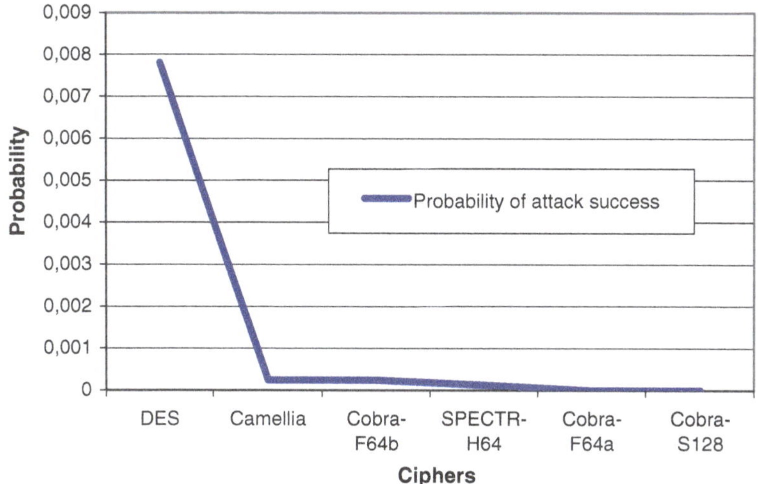

Fig. 4.4 Ciphers strength against attack success

These ciphers also can be comparing using the notion of the security margin (SM). It is one of the main security characteristics. Then percentage of *SM* more, then the cipher more vulnerable to attacks. SM can be estimated as

$$SM = \frac{100\,\%(R - R\text{min})}{R\text{min}} \tag{4.3}$$

where Rmin is the minimum number of round that is sufficient to provide security against differential analysis, R is nominal number of rounds. Rmin is defined by the block size, probability and number of rounds of the differential characteristic. For Cobra-S128, $SM = 50\,\%$ for Spectr-H64, $SM = 33\,\%$ for Camellia $SM = 50\,\%$ and for DES $SM = 77\,\%$. These results show that modified ciphers based on Feistel-scheme less vulnerability to attacks then DES or Camellia.

From the comparison done the results shows that there is a higher breakage probability on DES compared to our modified ciphers based on Feistel-scheme.

However from the graph we learn that there is less code breakage in modified Cobra-F64b, SPECTR-H64, Cobra-F64a and Cobra-S128 ciphers.

4.6 Summary

In this chapter we have presented an advance improved Feistel cipher based scheme which can be used in WSN block-cipher design for security by using CPB crypto primitives. Also we have shown how the new generation attacks are increasing with time, becoming complicated and mitigating against WSN and other fields respectively. Our analysis on comparison verified that there is less probability of code breakage in modified Feistel based scheme.

Our study argue that there is a benefit of using an improved Feistel scheme for WSN security, as its much easier to encrypt the data packet than encrypt data stream which most of the encryption standards are being used for at present. However Feistel scheme based can attain high and stable WSN security using block-ciphers compared to differential cryptanalysis. Due to Sensors efficiency in energy use, the modified ciphers are appropriate for their security design. This work saves as a notifications and mile stone to attract more attention to WSN security and DDP-based block-ciphers applications.

References

1. Mayank Saraogi. Security in wireless sensor networks. Computer and network security. http://www.cs.utk.edu/~saraogi/594paper.pdf (2006)
2. Karlof, C., Wagner, D.: Secure routing in wireless sensor networks: Attacks and countermeasures. In: Proceedings of the 1st IEEE International Workshop on Sensor Network Protocols and Applications, pp. 113–127 (2003)
3. Rasool, R.U., Guo, Q.P.: Security in Wireless Networks and Users-Grid. Course Work, Wuhan University of Technology (2004)
4. Mauw, S., van Vessem, I., Bos, B.: Forward secure communication in wireless sensor networks. In: Third International Conference Security in Pervasive Computing, pp. 32–42 (2006)
5. Hu, F., Ziobro, J., Tillett, J., Sharma, N.K.: Secure wireless sensor networks: problems and solutions. J. Syst. Cyber. Inf. **1**(9) (2004)
6. Kumar, S., Valdez, R., Gomez, O., Bose, S.: Survivability evaluation of wireless sensor network under DDoS attack. In: Proceedings of the International Conference on Networking, International Conference on Systems and International Conference on Mobile Communications and Learning Technologies, p. 82 (2006)
7. Bilstrup, U., Sjoberg, K., Svensson, B., Wiberg, P.A.: Capacity limitations in wireless sensor networks. In: Proceedings of the 9th IEEE International Conference on Emerging Technologies and Factory Automation, vol. 1, pp. 529–536 (2003)
8. Moldovyan, N.A., Moldovyan, A.A., Goots, N.D.: Variable bit permutations: linear characteristics and pure VPB-based cipher. Comput. Sci. J. Moldova **13**(1), 84 (2005)

9. Moldovyan, N.A., Moldovyanu, P.A., Summerville, D.H.: On software implementation of fast DDP-based ciphers. Int. J. Netw. Secur. **4**(1), 81–89 (2007)
10. Moldovyan, N.A., Moldovyan, A.A.: Data-driven ciphers for fast telecommunication systems, p. 202. Auerbach Publications. Taylor & Francis Group, New York, (2007)
11. Schneier, B.: Applied Cryptography: Protocols, Algorithms, and Source Code, 2nd edn, p. 758. Wiley, New York (1996)
12. Biham, E., Shamir, A.: Differential cryptanalysis of the full 16-round DES. In: Proceedings of the 12th Annual International Cryptology Conference on Advances in Cryptology, pp. 487–496 (1992)
13. Matsui, M.: Linear cryptanalysis method for DES cipher. Workshop on the Theory and Application of Cryptographic Techniques on Advances in Cryptology, pp. 386–397 (1994)
14. Levis, P.: Sensor network protocol design and implementation. Technical Report CS268, UC, Berkeley (2005)
15. Hill, J., Szewczyk, R., Woo, A., Hollar, S., Culler, D., Pister, K.: Proceedings of the Ninth International Conference on Architectural Support for Programming Languages and Operating Systems, vol. 35(11), pp. 93–104 (2000)
16. Feistel, H.: Cryptography and computer privacy. Sci. Am. **228**(5), 15–23 (1973)
17. Net and Electronic Frontier Foundations (EFF). RSA data security: RSA code-breaking contest again won by distributed. http://www.rsa.com/pressbox/html/990119-1.htm (1999)
18. Bodrov, A.V., Moldovyan, A.A., Moldovyanu, P.A.: DDP-based ciphers: differential analysis of SPECTR-H64. Comput. Sci. J. Moldova **13**(3), 268–291 (2005)
19. Moldovyan, N.A.: Fast DDP-based ciphers: design and differential analysis of Cobra-H64. Comput. Sci. J. Moldova **11**(3), 292–315 (2003)
20. Lu, J., Lee, C., Kim, J.: Related-key attacks on the full-round Cobra-F64a and Cobra-F64b. In: Proceedings of the Fifth Conference on Security and Cryptography for Networks, vol. 4116, pp. 95–110 (2006)
21. Goots, N.D., Moldovyan, A.A., Moldovyan, N.A.: Fast encryption algorithm SPECTR-H64. In: Proceedings of the International Workshop on Information Assurance in Computer Networks: Methods, Models, and Architectures for Network Security, pp. 275–286 (2001)
22. Keliher, L.: Toward provable security against differential and linear cryptanalysis for Camellia and related Ciphers. Int. J. Netw. Secur. **5**(2), 167–175 (2007)
23. Rudolf, D.: Optimized differential cryptanalysis of the data encryption standard. Department of Computer Science, University of Saskatchewan. http://www.cs.usask.ca/~dtr467/400/final/ (2001)

Chapter 5
The Distributed Signature Scheme (DSS) Based on RSA

5.1 Introduction

The Distributed Signature Scheme (DSS) is another important security service. It enables sensor nodes to communicate securely with each other. The main problem is to establish a secure signature between communicating nodes. However some special features (i.e. resource constraint, impracticalness of protecting or monitoring each individual node physically as well as their applications which usually being supported by many components such as routing and localization) of sensor networks make it particularly challenging to provide security services for sensor networks. This chapter describes a secret distribution scheme for sensor networks that achieves automatic secret redistribution. The goal is to support distributing the secret among new members joining a sensor network without involving a trusted agent or intervention from the user. Our analysis indicates that our new schemes have some nice features compared with the previous methods. In particular the system is efficient. Secondly, it guarantees automatic key distribution after initializations. Third it does not need urgent for key distribution and, finally, it automatically interact nodes coalition.

WSNs have recently emerged as an important means to study and interact with the physical world. A sensor network typically consists of a large number of tiny sensor nodes and possibly a few powerful control nodes (also called base stations). Many protocols and algorithms (e.g. routing, localization) will not work in hostile environments without security protection [1].

The use of aggregation in WSN allows to increase the efficiency and significantly survivability of sensor nodes. In this presentation many aspects of the WSN are being examined including security and efficient data aggregation [2–7]. For example, we will use Base Station (BS) to define the integral characteristic of any part of WSN; and assign one of the nodes as aggregator for clear elaboration and understanding of our presentation. The node will gather the needed information from the area, calculate the aggregation functions (i.e. average, min, max) and

© Springer International Publishing Switzerland 2016

G.S. Oreku and T. Pazynyuk, *Security in Wireless Sensor Networks*,
Risk Engineering, DOI 10.1007/978-3-319-21269-2_5

transfer this value to the BS. By so doing this will facilitate the cut down of total transmition cost rather than with the use of aggregator. But, all in all there are a need of special and reliable aggregation algorithms when it comes to node failing fulfilling their tasks. e.g., when the adversary can capture the nodes and change their functionality or the aggregator is compromised and brings total destruction to its function, i.e. when aggregator is compromised and sends the wrong information to the BS. For solving this kind of problems, special cryptography procedures can be used [8–13]. Some of the solutions sited might allow to BS to define incorrect aggregation result with high probability. And in this case the aggregation might be called reliable.

It's clear, that it's necessary to provide reliability requirements to transmit some extra data from aggregator to BS. In this case we argue that these data capacity (size) should be minimized with given reliability. In the existing reliability aggregation protocols at present, the size of extra data used is sufficiently high. This sets conditions and motivations for further interest in creating new reliable aggregation protocols, though it should be noted that creating special protocols for WSN also have some shortcomings; mainly being high number of keys to be kept by each sensor. However in providing reliable aggregation in WSN, the key management protocol issues should also be realized. The present solutions used for classical networks are unable to implement some of these options to WSN due to sensor's limitations and unfeasibility of using sensor's infrastructure.

Talkless of lots of constraints when it comes to providing security services in sensor networks, it also turns out to be a very challenging task. With the same lane of creating reliability in data aggregation, we introduce our finding to solve the problem of scheme distribution for signing the accurate information within nodes participating in transmitting the final information to BS. Having this kind of mechanism will allow to substantially decreasing energy consumption by eliminating the transmition of fake packets within the sensor networks meanwhile enhance the accuracy of security.

In this chapter we have presented distributed signature scheme design based on RSA (well-known encryption system using in a big amount of applications). Our scheme has three advantages. First, the mathematics presentations are provably secured. Secondly, the scheme is efficient, third together, we have proposed a secure, efficient proactive RSA based scheme with three security properties which did not exist in previously schemes.

5.2 RSA Based Secure Schemes

Blakley and Shamir invented secret sharing schemes independently. In Blakley's scheme [14], the intersection of *m of n* vector spaces yields a one-dimensional vector that corresponds to the secret. Wong et al. scheme [15] is one of several to catch a dealer that attempts to distribute invalid shares. Desmedt et al. [16] also present protocol to perform non-interactive verifiable secret redistribution

(VSR) that mitigates these problems in static sensor networks. VSR divides the sensor field into control groups each with a control node. Data exchange between nodes within a control group happens through the mediation of the control head which provides the common key. The keys are refreshed periodically and the control nodes are changed periodically to enhance security. SECOS enhances the survivability of the network by handling compromise and failures of control nodes. Gennaro et al. present a verification of a signature using a regular public key and a standard verification procedure; hence the verifier of a signature does not need to be aware of the form (centralized or distributed) in which the signature was generated, or who were the parties involved, nor does the signature increase in size as a function of the number of signers [17].

Our DSS scheme differs from previous VSR schemes in that it achieves automatic secret redistribution without the use of agent's .Also, unlike in VSR schemes, with signature setting actions node members can associate independently in our DSS. However secret key distribution protocol is un-interactive and doesn't require agent participation after scheme initialization.

Kong, et al. proposed a proactive RSA scheme for large-scale ad hoc networks [18, 19]. In their scheme, every node in ad hoc networks has a secret share of the secret key (the private key *d*). Nodes within one-hop distance jointly perform issuing certificates and refreshing their secret shares. The scheme is efficient. Unfortunately, the scheme has proved faulty [20, 21]. All the previous schemes [8, 9, 22] can be considered as special instances in this framework. Also Rui-shan et al. [23], have presented a new proactive RSA scheme for ad hoc networks, which includes four protocols, the initial key distribution protocol, the share refreshing protocol, the share distribution protocol, and the signature generation protocol. Their work mainly based on use of efficient proactive threshold RSA signature scheme. The initial key distribution protocol is used to distribute the initial secret shares to $2t + 1$ R nodes. Before distributing the secret key, they assume that a setup process has been carried out in which the RSA key generation took place and the RSA key pair has been computed where by in our work the agent is used to initialize the distributed signature's scheme and hence, all the remaining process is independently operated.

By instantiating the components in above frameworks, we further develop our DSS based on RSA with automatic signature setting procedure which provide coalition between (nodes) members, system with self-organizing property, i.e. the agent is not involved after initialization and during secret distribution process.

5.3 RSA Based Distributed Signature Scheme

Using only symmetric algorithms with authentication of sending data from sensors to BS have some disadvantages as well. e.g., only BS might be able to authenticate final report sent by aggregator towards BS. This means, there is a chance of any compromised node to be sent into network and by chance this fake packets might only be detected or thrown off by BS at the end point. With accomplishment of the all process the sensor node's resources would have been consumed for sending the fake packets.

For sensor networks with more powerful nodes, solving this kind of problem can be based on the use of distributive asymmetric signature. This sort of signature assumes the distribution of "digital signature of asymmetric algorithm secret key" by threshold circuit (scheme) key distributed between the all scheme members. Also this scheme assumes the presence of protocol which allows coalition from a given number of members to compute digital signature for given message in distributed manner. Regarding the fake packets filtering task in WSN the digital asymmetric signature algorithm can be used as follows.

Agent chooses and distributes digital signature's chosen algorithm secret key between all the sensors and hence, all the sensors are initialized by public key. For sending the aggregation result to BS, the results are signed by given number of sensors using signature distribution protocol. Furthermore each sensor, retransmitting data packet by using public key, can check the packet by itself and if the signature does not surpass the checking, it is automatically ejected from the network.

5.3.1 Distributed Signature Features

For effective working of system, distributed signature protocols should have some features:

- Independent work of the members during the initialization of signature. If the number of members is increased, this feature enables not to initialize this protocol again. Also it reduces signature setting delay.
- Self-organization, i.e. the system should be able to work automatically after initialization.
- Distribution of the new projection of secret should be non-interactive.

For the interactive protocol assuming the process of data exchange between working (nodes) members is an essential shortcoming due to limited traffic capacity existing in today's many WSN, additionally it increases energy consumption, whereas the synchronization in WSN is necessary. The schemes which can guarantee security are suitable for WSN security at present.

DSS with two features described above can easily be established based on El-Gammal or DSS digital signature [17, 24]. Unlike these digital signatures, based on RSA digital signature, no existing work, to the best of our knowledge, has addressed the issue of developing distributed signatures schemes with above-listed features at the moment. However, RSA digital signature has one important feature which does not exist in El-Gammal and DSS schemes. For RSA signature, the signature checking procedure is substantially accelerated if public key value is correctly chosen. This characteristic provides significant advantage for RSA distributed signature use in WSN.

The Scheme Definition. *The system model assumes that we have n nodes and one malicious node (note that for this example we will use only one malicious node though in really application our approach should be able to withstand up to t − 1 compromised nodes). Also the system has a trusted agent which initializes the scheme. For this case agent chooses RSA secret key and distribute this key safely between the nodes providing (t, n)—threshold scheme. After this initialization the participation of trusted agent is not needed. Assuming that malicious user can compromise s < t of nodes and since malicious is able to break the multiple signature scheme, he could execute attack by chosen message (we call it chosen-message attack, CMA); therefore he could request any of n nodes members to invoke signature protocol for any chosen message. In this situation malicious user aim is either tamper message signature which he/she did sign or disrupt a wrong message signed by another member.*

5.3.2 RSA Based Secret Key Distributions Main Approaches

In existing works based on distributed RSA signature there are three main approaches as far as RSA based secret key distribution is concerned.

- In the first approach, the secret key *d* is distributed according to Shamir secret scheme distribution. The system working according to this approach is impossible without trusted agent participation.
- In the second approach, the level in secret scheme distribution has been added. Firstly, the secret is distributed between *n* nodes additively, and then every received projection is distributed by threshold circuit. Such schemes are interactive and are unable to work without agent.
- In the third approach, there is secret key *d* distributed. But this kind of separation brings vulnerability to distributed scheme. Moreover this approach doesn't assume independent working of members' nodes coalition during signature establishment.

Thus each of these approaches noticed above have some functional limitation and do not employ at least one of the features formulated above.

The scheme of secret distribution (sharing) is one of the DSS components. In particular, Shamir scheme can be used as scheme of secret distribution. In Shamir scheme there is polynomial function $f(x)$ as:

$$\mathrm{f}(x) = \left(\mathrm{f}_0 + f_1 x + \cdots + f_{t-1} x^{t-1}\right) \mathrm{mod\_}P \tag{5.1}$$

in which $f_0 = S$—secret, i.e. secret key, $f_1, \ldots, f_{t-1}$—random values, t—number of secret's projections (sub-keys) or number of coalition's members and P—prime number. Each member of protocol gets the secret projection as $ss = f(id)$, where id

is member's ID. Any coalition K of t members could restore (recover) the secret $f_0 = f(0)$ using Lagrange interpolation

$$f(0) = \sum_{u \in K} ss_u l_u(0) (\bmod P) \tag{5.2}$$

where $l_u(x)$ are Lagrange coefficients.

For Shamir scheme to be used in distributed RSA signature, it's necessary to choose the secret S and module P. For distributed RSA signature secret key d is the secret. Relatively to module P there are two ways, either make it public, e.g. $P = N$ or make it secret, i.e. $P = \phi(N)$ or $P = \lambda(N)$. If P *is* known (e.g. RSA module N), that brings information leakage and interdependency of coalition members actions during the distribution of signature setting procedure running. If P is a secret value (e.g. $P = \lambda(N)$ or $P = \phi(N)$), that gives the system possibility of being not self-organized according to the next statement.

Statement 1 In case of using module P and this P module is unknown to members, then it's necessary to have trusted agent for secure project distribution to new member.

It is necessary to ignore the use of module P approach to eliminate disadvantaged listed above. However, projection distribution procedure without agent remains interactive. In addition, abandonment of P increases projections size eventually to complexity of signature setting. It's easy to get higher estimation of secret projection size *(R)* by using:

$$R \leq \log(N) + (t - 1)k + 1 \tag{5.3}$$

where N—RSA module, t—coalition size (number of members), k—user ID length. For example, for *user ID length k = 48* bit and *coalition size t = 10*, *projection size* R will not be over $\log(N) + 48(t - 1) + 1 \approx 1500$ *bit* which means there is an increase of signature setting complexity of approximately 1.5 times.

5.4 Our Approach on Scheme Establishment

We propose to modify secret distribution scheme by getting rid of interaction in distribution procedure of new projection without agent (statement 1) and reduce the size of secret projection. We put into consideration prime number $Q > \max(id_i)$. We estimate the secret projection *f(x, y)* as follow:

$$f(x, y) = \sum_{i=0}^{t-1} \sum_{j=0}^{t-1} f_{i,j} (x^i \bmod Q)(y^j \bmod Q) \tag{5.4}$$

It's impossible to use LaGrange interpolation with such kind of secret distribution function. Instead, it's essential to solve the combined linear equations.

For secret recovering, each of coalition member node u calculate its function value with $x = 0$ getting

$$f(0, id_u) = f_0 + f_1(y \bmod Q) + \cdots + f_t(y^t \bmod Q) \tag{5.5}$$

with $y = id_u$. And $f_0 = f_{0,0}$. Having t values of given function, secret can be recovered by solving the following combined equations:

$$\begin{bmatrix} f(x_{i_1}) \\ f(x_{i_2}) \\ \cdots \\ f(x_{i_t}) \end{bmatrix} = G \begin{bmatrix} f_0 \\ f_1 \\ \cdots \\ f_{t-1} \end{bmatrix},$$

where

$$G = \begin{bmatrix} (x_{i_1})^0 \bmod Q & (x_{i_1})^1 \bmod Q & \cdots & (x_{i_1})^{t-1} \bmod Q \\ (x_{i_2})^0 \bmod Q & (x_{i_2})^1 \bmod Q & \cdots & (x_{i_2})^{t-1} \bmod Q \\ & & \cdots & \\ (x_{i_t})^0 \bmod Q & (x_{i_t})^1 \bmod Q & \cdots & (x_{i_t})^{t-1} \bmod Q \end{bmatrix} \tag{5.6}$$

Each coalition member calculates its function value with $x = id_{new}$ getting

$$f(id_{new}, id_u) = s_{new}(id_u) = s_0 + s_1(y \bmod Q) + \cdots + s_{t-1}(y^{t-1} \bmod Q) \tag{5.7}$$

with $y = id_u$ for projection to be distributed to new members without agent. Secret projection $(s_0, s_1, \ldots, s_{t-1})$ for new user can be calculated from the following combined equations:

$$\begin{bmatrix} s_{new}(x_{i_1}) \\ s_{new}(x_{i_2}) \\ \cdots \\ s_{new}(x_{i_t}) \end{bmatrix} = \begin{bmatrix} s_0 \\ s_1 \\ \cdots \\ s_{t-1} \end{bmatrix} \tag{5.8}$$

The following statement is true for proposed secret distribution scheme according to statement number 2.

Statement 2 For modified secret distribution scheme the following are true:

(i) Scheme has threshold (t) and it is safe;
(ii) The procedure of projection distribution in scheme is non-interactive and doesn't request agent participation;
(iii) The procedure of projection distribution is safe;
(iv) Size of each projection is no larger than $\log(N) + k + t$.

New RSA DSS based on proposed modified secret distribution scheme includes three steps:

5.4.1 Scheme Initialization

Agent generates the prime number $Q > \max(id_i)$.

(a) Agent generates public RSA key $N = pq$ and $e > Q$, where e—prime number, p and q are random prime numbers. Then agent generates secret key d as following: $ed = 1 \bmod \Phi(N)$.
$\Phi(N) = (p-1)(q-1)$, where e and d are public and closes parts of key. (N, e)—public key's part, d—closed.
(b) Agent generates function

$$f(x,y) = \sum_{i=0}^{t-1}\sum_{j=0}^{t-1} f_{i,j}(x^i \bmod Q)(y^j \bmod Q) \tag{5.9}$$

where $f_{0,0} = d$, and coefficients $f_{i,j} \in Z_N$ (Z_N is a set of prime numbers) was randomly chosen with $f_{i,j} = f_{j,i}$ condition.
(c) Each node u gets the function $s_u(x) = f(x, id_u)$ as a secret key projection.

5.4.2 Generation of Distributive Signature

(a) The coalition K of t members is chosen (cluster). Each node calculates partial signature by the equation $S_u(m) = m^{s_u(0)} \bmod N$, where m is hash-function value of signing message, $u \in K$.
(b) After getting t partial signatures, signature's collector makes the matrix G for coalition K members and reverses it over the rational number field.
(c) Signature collector calculates $G' = \lambda G^{-1}$, where λ is the least common multiple of all elements of matrix G^{-1}. Then signature collector calculates

$$S'(m) = \left(\prod_{j=1}^{t} \left(S_{u_j}(m)\right)^{g'_{1j}}\right) \bmod N \tag{5.10}$$

(d) Using extended Euclid algorithm, collector finds x and y from $x\lambda + ye = 1$.
(e) Calculating of the signature as

$$S(m) = ((S'(m))^x m^y) \bmod N. \tag{5.11}$$

5.4.3 *Key Projection Distribution to New User*

(a) For getting secret key projection the new node u has to find coalition K from t as already initialized nodes and report them to its own id_{new}.
(b) Every coalition member u calculate its own function value with $x = id_{new}$, getting

$$f(id_{new}, id_u) = s_{new}(id_u) = s_0 + s_1(y \bmod Q) + \cdots + s_{t-1}(y^{t-1} \bmod Q) \tag{5.12}$$

With $y = id_u$.
(c) New node finds its secret projection $(s_0, s_1, \ldots, s_{t-1})$ from combined of linear equations:

$$\begin{bmatrix} s_{new}(x_{i_1}) \\ s_{new}(x_{i_2}) \\ \cdot \\ s_{new}(x_{i_t}) \end{bmatrix} = G \begin{bmatrix} s_0 \\ s_1 \\ \cdot \\ s_{t-1} \end{bmatrix} \tag{5.13}$$

Thus, secret key projection distribution to a new node does not request the agent participation and it is not interactive.

For proposed scheme it can be seen that it allows generating correct RSA based signature, and the next statement is true.

Statement 3 Proposed distributed signing scheme provides high security guarantee and even safer as RSA.

5.5 Summary

In this chapter we developed a RSA based distributed signature scheme with independent member nodes behavior, signature signing setting and un-interactive projection distribution protocol secret key with no agent participation. The proposed distributed signature scheme unlike existing schemes has the following advantages:

- Nodes in cluster can associate independently during the signature distributing;
- Secret key distribution protocol is non-interactive and doesn't require agent participation after scheme initialization.

As one of the possible future directions, we observed that sensor nodes have low mobility in many applications. Thus it may be desirable to develop location- based schemes so that the nodes can directly establish a signature setting automatically.

References

1. Liu, D., Ning, P.: Security for wireless sensor networks. Advances in Information Security, p. 28. Springer Science and Business Media, Berlin (2007)
2. Zhu, H., Bao, F., Deng, R.H., Kim, K.: Computing of trust in wireless networks. In: Proceedings of 60th IEEE Vehicular Technology Conference, vol. 4, pp. 2621–2624 (2004)
3. Estrin, D., Govindan, R., Heidemann, J.S., Kumar, S.: Next century challenges: Scalable coordination in sensor networks. In: Proceedings of the 5th Annual ACM/IEEE International Conference on Mobile Computing and Networking, pp. 263–270 (1999)
4. Hu, L., Evans, D.: Secure aggregation for wireless networks. In: Proceedings of the 2003 Symposium on Applications and the Internet Workshops, pp. 384–391 (2003)
5. Madden, S., Franklin, M.J., Hellerstein, J.M., Hong, W.: Tag: A tiny aggregation service for ad-hoc sensor networks. ACM SIGOPS Oper. Syst. Rev. **36**(SI), 131–146
6. Przydatek, B., Song, D., Perrig, A.: Sia: Secure information aggregation in sensor networks. In: Proceedings of the 1st International Conference on Embedded Networked Sensor Systems, pp. 255–265 (2003)
7. Shrivastava, N., Buragohain, C., Agrawal, D., Suri, S.: Medians and beyond: New aggregation techniques for sensor networks. In: SenSys '04: Proceedings of the 2nd International Conference on Embedded Networked Sensor Systems, pp. 239–249 (2004)
8. Eschenauer, L., Gligor, V.D.: A key-management scheme for distributed sensor networks. In: Proceedings of the 9th ACM Conference on Computer and Communications Security, pp. 41–47 (2002)
9. Chan, H., Perrig, A., Song, D.: Random key predistribution schemes for sensor networks. In: Proceedings of the 2003 IEEE Symposium on Security and Privacy, vol. 197(23), pp. 197–213 (2003)
10. Hwang, J., Kim, Y.: Revisiting random key pre-distribution schemes for wireless sensor networks. In: Proceedings of the 2nd ACM Workshop on Security of Ad hoc and Sensor Networks (SASN '04), pp. 43–52 (2004)
11. Liu, D., Ning, P., Li, R.: Establishing pairwise keys in distributed sensor networks. ACM Trans. Inf. Syst. Secur. **8**(1), 41–77 (2005)
12. Schneier, B.: Applied Cryptography: Protocols, Algorithms, and Source Code (2nd ed). Wiley, New York, p. 758 (1996)
13. Gura, N., Patel, A., Wander, A., Eberle, H., Shantz, S.: Comparing elliptic curve cryptography and RSA on 8-bit cpus. In: 2004 Workshop on Cryptographic Hardware and Embedded Systems, pp. 119–132 (2004)
14. Gennaro, R., Halevi, S., Krawczyk, H., Rabin, T.: Threshold RSA for dynamic and ad-hoc groups. In: Advances in Cryptology—EUROCRYPT 2008, 4965/2008, pp. 88–107. Springer, Berlin
15. Wong, T.M., Wang, C., Wing, J.M.: Verifiable secret redistribution for archive systems. In: Proceedings of the First International IEEE Security in Storage Workshop, pp. 94–105 (2002)
16. Desmedt, Y., Jajodia, S.: Redistributing secret shares to new access structures and its applications, Technical Report ISSE TR-97-01, George Mason University (1997)
17. National Institute of Standards and Technology.: Digital signature standard. JIST FIPS PUB 186, U.S. Department of Commerce (1994)
18. Jie-jun, K., Zerfos, P., Hai-yun, L., Song-wu, L., Li-xia, Z.: Providing robust and ubiquitous security support for MANET. In: IEEE 9th International Conference on Network Protocols (ICNP), p. 251 (2001)
19. Hai-yun, L., Jie-jun, K., Zerfos, P., Song-wu, L., Li-xia, Z.: URSA: Ubiquitous and robust access control for mobile ad hoc networks. In: IEEE/ACM Transactions on Networking (ToN), vol. 12(6), pp. 1049–1063 (2004)
20. Narasimha, M., Tsudik, G., Yi, J.H.: On the utility of distributed cryptography in P2P and MANETs: The case of membership control. In: IEEE 11th International Conference on Network Protocol (ICNP), pp. 336–345 (2003)

21. Jarecki, S., Saxena, N., Yi, J.H.: Cryptanalyzing the proactive RSA signature scheme in the URSA ad hoc network access control protocol. In: ACM Workshop on Security of Ad Hoc and Sensor Networks (SASN) (2004)
22. Blakely, G.R.: Safeguarding cryptographic keys. In: Proceedings of the National Computer Conference, vol. 48 (1978)
23. Rui-shan, Z., Ke-fei, C.: An efficient proactive RSA scheme for large-scale ad hoc networks. J. Shanghai Univ. Engl. Ed. **11**(1), 64–67 (2007)
24. ElGamal, T.: A public key cryptosystem and a signature scheme based on discrete logarithms. IEEE Trans. Inf. Theory. **IT-31**(4), 469–472 (1985)

Chapter 6
Reliable Data Aggregation Protocol for Wireless Sensor Networks

6.1 Introduction

Current routing protocols in WSNs or even in Wireless Ad hoc Networks are very susceptible to many attacks i.e. stealthy attack. The most simple among those is where the adversary injects malicious routing information into the network. This results in routing inconsistencies leading to high increase in end-to-end delays or even packet losses in the network. In this chapter, first, we abstract two fundamental routing protocols, which can be generally grouped in two broad categories based on the intrinsic nature of WSN. We argue that none of previous proposed routing protocols satisfies all of them at the same time.

The novelty of our protocol is building a general routing protocol based on two methods, which takes into consideration two factors: addition of sensor nodes to the aggregation process and by considering complex report interaction between base station and aggregator. Finally, to evaluate the efficiency of proposed protocol, we carry out comparison experiment of our proposed protocol to general known protocols. Performance cost evaluation of our proposed protocol shows essential advantage over existing protocol.

Routing protocols can be generally grouped in two broad categories: reactive and proactive protocols. Proactive routing protocols use some kind of periodic beaconing or coordination mechanisms between nodes to pro-actively maintain routing tables at each node. Conversely, reactive protocols don't attempt to maintain routing tables continually; in-stead, they initiate route discovery only when the route is required for a packet transmission. Discovered routes are temporarily cached to be used for subsequent requests addressed to the same node, but will eventually expire after a period of inactivity.

Attacks against the routing protocols generally involve the manipulation of malicious users of route messages and injection of false or incomplete information. Routing table overflow, cache poisoning, and network flooding are also possible and relatively simple kinds of attacks.

© Springer International Publishing Switzerland 2016

G.S. Oreku and T. Pazynyuk, *Security in Wireless Sensor Networks*,
Risk Engineering, DOI 10.1007/978-3-319-21269-2_6

An increasing number of advanced attacks at the routing layer includes, for instance, the creation of wormholes to tunnel traffic through an undisclosed hidden path without the knowledge of source or destination nodes, or blackhole attacks in which a malicious node falsifies route advertisements to misdirect the traffic addressed to a victim, consuming the traffic without forwarding it.

Numerous proposed countermeasures aim to prevent routing attacks, ranging from encryption at the protocol level to information correlation between multiple nodes to packet leash protocols. However, significant opportunity remains for further work in the area.

Node and message authentication are also critical issues that span multiple levels of the protocol stack. Authenticating nodes in Manets generally follow the same cryptographic strategies used for authentication in wired networks. The basic challenge lies in adapting key management protocols, which are the basic part of any security communications infrastructure. Although key management is still an open and active area of research, several researchers have proposed strategies for distributed key management, leveraging for instance, for threshold cryptography strategies [1], dynamic cluster-based certificate of authorities [2] and fully distributed schemas based on certificate chaining for public key authentications. For mission-specific applications, tactical network nodes are often configured with time-sensitive pre-shared keys, built as a part of standard system images created for different missions.

Often WSNs are developed in the open and readily available territory; therefore, it is always necessary to use special procedures to protect transferred information against possible casual or deliberate distortions. Protection of transferred information against deliberate distortions requires well-developed authentication methods in the whole process of messages transportation. For this purpose, techniques like messages authenticity checking codes (MAC-codes) are used.

There are two types of data aggregation protocols with saving of reliability:

The first one is based on distribution of aggregation process by involving additional sensor nodes.
The second one is based on complex report interaction between base station and aggregator.

According to the first type, aggregator should prove to base station a correctness of the presented result. As per the first approach, the scheme is based on a treelike routing, where the tree root is supposed to be the base station. The aggregation routing is supposed to be from twigs to root which according to our presentation is the base station (BS). Thus, modular function is calculated from values received from descendants in each unit, and the calculated value together with arguments values are transferred to the parent-node. In this case the parent-unit can check correctness of data aggregated in affiliated nodes. However, the given scheme does not possess sufficient reliability; in particular, the aggregation result sometimes can appear incorrect due to misoperation of two subsequently nodes in a tree. Moreover these approach had some significant limitation i.e. the number of calculated aggregation functions are limited.

Another approach based on the use of so-called witness nodes that actually duplicate aggregator action. By matching the witness's and received aggregation results the signing is authenticated and aggregator sends the result and the witness's signatures to BS.

The disadvantage of the presented mechanism is that the volume of data transferred by control sensor increases linearly according to witness's nodes number. Using this approach, there is a following protocol: aggregator collects the data from sensors, calculates the aggregated value, signs it and sends to BS. Subsequently, the interactive protocol of calculations proof carries out between aggregator and BS. The disadvantage of this approach is a huge data volume transferring between aggregator and BS.

To avoid the problems mentioned above and to achieve reliable data aggregation in WSN we propose the new protocol bases on distributed verification of aggregation result.

6.2 Problem Statement

Routing protocols are highly susceptible to node capture attacks. It is observed and analyzed that even a single node capture is sufficient for an attacker to take over the entire network. Unlike traditional networks, where physical security can prevent such conditions, WSN belongs to extremely hostile and unattended environments.

Usually, a WSN consists of a large number of sensor nodes which are deployed in some area distant from the home server. These sensor nodes perform measurements and route the information towards the BS. However, in order to save the communication bandwidth, these readings are aggregated at intermediate points in the network which are called as aggregators. Some sensor networks have a single aggregator, which is usually the BS itself or as others [3] have several aggregators where each non-leaf node is an aggregator, as also shown in Fig. 6.1.

In this setting, there are two major attacks over the information being aggregated [4]. First is the stealthy attack, where the attacker's goal is to make the home server accept false aggregation results, which are very much different from the actual

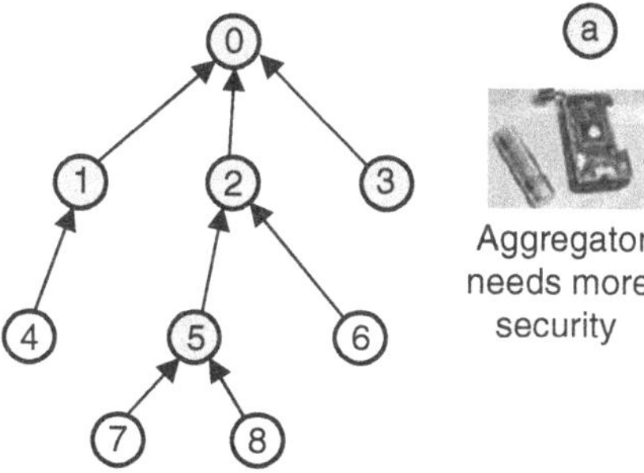

Fig. 6.1 Secure group management. Aggregators and base stations

results determined by the measured values. Moreover, the attacker also wishes that the homing server is not able to detect these changes. So he does not launch a DoS attack by not responding with the aggregated values at all.

6.3 Existing Data Aggregation Protocols

In typical WSN, sensor nodes are usually resource-constrained and battery-limited. In order to save resources and energy, data must be aggregated to avoid overwhelming amounts of traffic in the network. There has been extensive work on data aggregation schemes in sensor networks [5–8]. There are some works [9, 10] which investigates secure data aggregation schemes in the face of adversaries who try to tamper with nodes or steal the information.

Also a dependable and efficient data aggregation scheme based on fault map that is constructed by estimated fault probability using Bayesian Belief Network (BBN) has been proposed [11], or, other authors presented two privacy-preserving data aggregation schemes, Cluster-based Private Data Aggregation (CPDA)–leverages clustering protocol and algebraic properties of polynomials and Slice-Mix-AggRegaTe (SMART)–builds on slicing techniques and the associative property of addition [12].

Most research efforts in this area are directed to the development of new protocols that promote efficient resources utilization, mainly with respect to the power consumption [13, 14]. The protocol based on the concept of delayed aggregation peer monitoring and requiring local interactions only [15] proposes to provide both confidentiality and integrity of the aggregated data, to detect bogus data injection attempts, and to provide high resilience to sensor failures.

The LEACH protocol [16] is a hierarchical self-organized cluster based approach for monitoring applications. Al-Karaki et al. [17] propose exact and approximate algorithms to find the minimum number of aggregation points in order to maximize the network lifetime. This paper [17] does not justify that the optimal selection remains the same along the network lifetime. This proposal resolves the problems of routing and data aggregation as one joint problem.

Patil et al. [18] use the ability of space-filling curves to index the sensor nodes and Krisnamachari et al. [19] examine the complexity of optimal data aggregation, showing that although it is an NPhard problem, there are useful polynomial-time special cases. Lindsey et al. propose PEGASIS [13], an extension of LEACH, where nodes can transmit to any other node of the system and to the BS. Nodes transmit to their nearest neighbor and messages are transmitted to the BS on rotation basis. They are organized to form a chain, which can be computed in a centralized way by the BS and can broadcast to all nodes or is controlled by the sensor nodes themselves using a greedy algorithm [19] which resolves both the problems of routing and data aggregation.

6.4 Proposed Protocol

We assume that there are following participants in our proposed data aggregation protocol: base station (BS), aggregator (A), sensor nodes (S_j) and t—number of verifier-nodes (V_i). Aggregator and verifier-nodes are the typical sensor nodes chosen randomly by base station inside the clusters. Periodically, BS reassigns the aggregator and verifiers.

The protocol consists of three stages:

- aggregation result calculation,
- checking the received result by t of nodes-verifiers,
- sending aggregation result together with verifier's signatures to the BS.

At the first stage all sensors send their data to aggregator:

$$S_j \to A : \{data_j,\ MAC(K_{S_j,A},\ data_j)\} \tag{6.1}$$

where $data_j$ is the data changed by sensor S_j, $K_{Sj,\,A}$ is a common key of sensor S_j and aggregator A, MAC is a message authentication code. Aggregator collects all data, checks their authenticity, using corresponding MAC, and calculates aggregation result.

At the second, there is a checking of aggregated result.
This stage consists of two steps.
Aggregator sends all the collected data to verifiers:

$$A \to Verifiers : \{D,\ MAC(K_{V_1,A},\ D), \ldots, MAC(K_{V_t,A},\ D)\} \tag{6.2}$$

where

$$D = \{data_1, \ldots, data_n\} \tag{6.3}$$

and $K_{Vi,\,A}$ is a common key of verifier V_i and aggregator A.

Each verifier $V_i(i = \overline{1,t})$ analyzes corresponding authentication code in the received package. If MAC is correct, the verifier randomly chooses k sensors and requests data from them. Each sensor S_j, having received query from V_i, sends the answer:

$$S_j \to V_i : \{data_j,\ MAC(K_{S_j,V_i},\ data_j)\} \tag{6.4}$$

where $K_{Sj,\,Vi}$ is a common key of sensor node S_j and verifier V_i, $data_j$ is a data of S_j.

After verifier V_i has checked up an authenticity of MAC from sensor S_j, it compares the corresponding data received from aggregator and sensor. If these values differ, the verifier sends warning message to the base station. If, during verification, node V_i does not find data inconsistency, it signs aggregation result generally with the BS key:

$$V_i \to A : \{SN_{V_i},\ MAC(K_{V_i,A},\ SN_{V_i})\} \tag{6.5}$$

where

$$SN_{V_i} = MAC(K_{V_i,BS},\ AR) \tag{6.6}$$

AR is aggregation result and $K_{Vi,\ BS}$ is a common key of verifier V_i and BS.

At the third stage aggregator collects signatures from all nodes-verifiers, forms the report, signs it and sends to the base station:

$$A \to BS : \{AR,\ SN_{V_1} \oplus \ldots \oplus SN_{V_t} \oplus MAC(K_{A,BS},\ AR)\} \tag{6.7}$$

where $K_{A,\ BS}$ is a common key of aggregator A and BS.

For checking received report the base station calculates all signatures, unites them, using operation XOR, and compares their calculated value with received value. If there are no differences, the result is accepted and considered as correct.

For the given protocol it is possible to calculate receiving distortion probability of aggregation result by the BS. We calculate it as follows:

$$\mathrm{P}_{er} = \left(\frac{C_{n-m}^k}{C_n^k} + \left(1 - \frac{C_{n-m}^k}{C_n^k}\right)p\right)^t \tag{6.8}$$

where n—quantity of sensors in cluster, t—quantity of nodes-verifiers, k—quantity of queries from each node-verifier, m—quantity of the distortion reports in a package, given to verifiers for checking, P—probability of incorrect work of verifier node.

6.5 Security Assumptions

In the case of very low efficiency of the additional sensors, the limitations on possible security services are very significant. However, appropriate use of our secure protocol for sensor networks can provide such security services as: system availability, authorization of sensors, confidentiality of transmitted information, and freshness and integrity of the measured data.

6.6 Experiment Evaluation

We use simulation written in C++ to investigate the effect of the various parameters on different Distributed Sensor Networks (DSN) sizes. Of particular interest are the efficiency and scalability of our scheme and also the determination of some

Table 6.1 Comparison of communication costs of reliable aggregation in various protocols

Costs type/protocol	SIA	Witnesses	Our protocol
Sensor transferring (byte)	22	42	36
Transferring out of cluster (byte)	922	22	22
Total amount of data transferred within cluster (byte)	2200	4230	3916
Total amount of data received within cluster (byte)	2200	16,830	4432
Maximum error of receiving aggregation result (%)	40	5	5

parameter values cannot be easily computed, such as the diameter of the resulting secure network. The simulations assume a network of 1000 nodes with an average density of 40 sensor nodes in a neighborhood. Each simulation is run 10 times with different seeds for the random generator, and the results represent the average values on the 10 runs, unless otherwise noted.

In Table 6.1 we compare the communication costs of proposed protocol and existing known protocols.

Our comparison was made based on hundred nodes cluster, with 0.1 probability of incorrect work of verifier node and with probability of aggregation distortion result received by base station, being less or equal to 0.05. In a package sent by a sensor node, data takes two bytes, and authentication code takes 10 bytes. Communication comparison cost shows essential advantage of our proposed protocol compared to known protocol.

6.7 Summary

The research stream has evolved beyond the original conception of data transferring for routing protocol based on the data aggregation that can satisfy the communication cost requirements which is one of the most fruitful research areas in the field of WSN security, but most of the extensions have evolved the pre and initial acceptance phases.

Our study extends understanding of inconsistencies in data aggregation which leads to high increase in end-to-end delays or even packet losses in the network especially in WSN. Results indicate that our proposed protocol is indeed having incomparable records in received and being transferred data amount, in and out of cluster, however it comprises minimal error aggregations upon receiving.

Our future research avenue is to investigate our finding and incorporate our protocol to secret automatic scheme for sensor networks in order to achieve automatic secret redistribution.

References

1. Shamir, A.: How to share a secret. Commun. ACM **22**(11), 612–613 (1979)
2. Hahn, G.: Cluster-based certificate chain for mobile Ad Hoc networks. In: Proceedings of the Compter Science and Its Application, pp. 769–778 (2006)
3. Madden, S., Franklin, M.J., Hellerstein, J.M., Hong, W.: Tag: a tiny aggregation service for ad-hoc sensor networks. SIGOPS Oper. Syst. Rev. **36**(SI), 131–146 (2002)
4. Przydatek, B., Song, D., Perrig, A.: Sia: secure information aggregation in sensor networks. In: Proceedings of the 1st International Conference on Embedded Networked Sensor Systems, pp. 255–265 (2003)
5. Tang, X., Xu, J.: Extending network lifetime for precision constrained data aggregation in wireless sensor networks. INFOCOM, 1–12 (2006)
6. Goyeneche, M., Villadangos, J., Astrain, J.J., Prieto, M., Cordoba, A.: A distributed data gathering algorithm for wireless sensor networks with uniform architecture. In: Proceedings of the 15th Euromicro International Conference on Parallel, Distributed and Network-Based Processing, pp. 373–380 (2007)
7. Al-Yasiri, A., Sunley, A.: Data aggregation in wireless sensor networks using the SOAP protocol. Sens. Appl. XIV, 1–6 (2007)
8. Hu, Y., Nuo, Y., Jia, X.: Energy efficient real-time data aggregation in wireless sensor networks. In: International Conference on Communications and Mobile Computing, pp. 803–808 (2006)
9. Yang, Y., Wang, X., Zhu, S., Cao, G.: SDAP: a secure hop-by-hop data aggregation protocol for sensor networks. ACM Trans. Inf. Syst. Secur. (TISSEC) **11**(4), 18 (2008)
10. Wagner, D.: Resilient aggregation in sensor networks. In: Proceedings of the 2nd ACM Workshop on Security of Ad Hoc and Sensor Networks, pp. 78–87 (2005)
11. Chang, Y.S., Huang, J.H., Juang, T.Y.: Dependable data aggregation on cluster-based wireless sensor networks. In: Proceedings of the 11th WSEAS International Conference on Communications, vol. 11, pp. 300–305 (2007)
12. He, W., Liu, X., Nguyen, H., Nahrstedt, K., Abdelzaher, T.: PDA: privacy-preserving data aggregation in wireless sensor networks. In: 26th IEEE International Conference on Computer Communications, pp. 2045–2053 (2007)
13. Lindsey, S., Raghavendra, C.S.: PEGASIS: power efficient gathering in sensor information systems. In: Proceedings of the IEEE Aerospace Conference, vol. 3, pp. 1126–1130 (2002)
14. Younis, M., Youssef, M., Arisha, K.: Energy-aware routing in cluster-based sensor networks. In: Proceedings of the 10th IEEE/ACM International Symposium on Modelling, Analysis and Simulation of Computer and Telecommunication Systems, pp. 129–136 (2002)
15. Di Pietro, Roberto.: Confidentiality and integrity for data aggregation in WSN using peer monitoring. Wiley, New Jersey (2009). (To appear in Security and Communication Networks Journal)
16. Heinzelman, W., Chandrakasan, A., Balakrishnan, H.: Energy-efficient communication protocol for wireless microsensor networks. In: Hawaii International Conference on System Science, vol. 8, p. 8020 (2000)
17. Al-Karaki, J., Ul-Mustafa, R., Kamal, A.: Data aggregation in wireless sensor networks—exact and approximate algorithms. In: IEEE Workshop High Performance Switching and Routing, pp. 241–245 (2004)
18. Patil, S., Das, S.: Serial data aggregation using space-filling curves in wireless sensor networks. In: 1st International Conference on Embedded Networked Sensor Systems, pp. 326–327 (2003)
19. Krishnamachari, B., Estrin, D., Wicker, S.: Impact of data aggregation in wireless sensor networks. In: DEBS'02: International Workshop on Distributed Event-Based Systems, pp. 575–578 (2002)

Conclusions and Future Work

WSNs are exposed to numerous security threats that can endanger the success of applications. Security support in WSNs is challenging due to the limited energy, communication bandwidth, and computational power. Also, sensors are often deployed in an open environment where no physical security is available. Given the diversity of WSN applications and possibly different security requirements, we think special enhancing security approaches to securing WSN is necessary. To recapitulate, the contributions of this book are.

Security of WSN is considered through QoS. Using QoS components, we evaluated models and system-level test using sensor nodes. One primordial issue is to satisfy application QoS requirements while providing a high-level abstraction that addresses WSN security. With the proposed approach, such tests can be easily parallelized by applying wireless broadcast to many nodes at once. As a result, the proposed approach can be used in variety of testing scenarios. A secure model is proposed with flow of security classes providing different levels of security using QoS. However our finding found that effects of security metrics place a lot of burden on the QoS of the overall system thus decreasing performance.

Mathematical model for WSN security. We need a theoretical foundation to determine the minimum number of sensors to be deployed so that intruders crossing a barrier of sensors will always be detected. However the sensor nodes deployed should have the security implemented in them as suggested with our findings. We present the fundamental mathematical model design for sensor nodes that can be used to secure different WSNs topology against intruders.

An advance improved Feistel cipher based scheme which can be used in WSN block-cipher design for security by using CPB crypto primitives. Also we have shown how the new generation attacks are increasing with time, becoming complicated and mitigating against WSN and other fields respectively.

Our study argue that there is a benefit of using an improved Feistel scheme for WSN security, as its much easier to encrypt the data packet than encrypt data stream which most of the encryption standards are being used for at present. However Feistel scheme based can attain high and stable WSN security using block-ciphers compared to differential cryptanalysis. Due to Sensors efficiency in energy use, the modified ciphers are appropriate for their security design. This work saves as a

© Springer International Publishing Switzerland 2016

G.S. Oreku and T. Pazynyuk, *Security in Wireless Sensor Networks*,
Risk Engineering, DOI 10.1007/978-3-319-21269-2

notifications and mile stone to attract more attention to WSN security and DDP-based block-ciphers applications.

RSA based distributed signature scheme was developed with independent member nodes behavior, signature signing setting and un-interactive projection distribution protocol secret key with no agent participation.

The proposed distributed signature scheme unlike existing schemes has the following advantages:

- with signature setting actions node members can associate independently.
- secret key distribution protocol is un-interactive and doesn't require agent participation after scheme initialization.

Reliable data aggregation protocol. The research stream has evolved beyond the original conception of data transferring for routing protocol based on the data aggregation that can satisfy the communication cost requirements which is one of the most fruitful research areas in the field of WSN security, but most of the extensions have evolved the pre and initial acceptance phases.

Our study extends understanding of inconsistencies in data aggregation which leads to high increase in end-to-end delays or even packet losses in the network especially in WSN. Results indicate that our proposed protocol is indeed having incomparable records in received and being transferred data amount, in and out of cluster, however it comprises minimal error aggregations upon receiving.

As WSNs continue to grow and become more common, we expect that further expectations of security will be required of these WSN applications. We hope that our research and contributions will likely make strong security a more realistic expectation in the future. We also expect that the current and future work in privacy and trust will make WSNs a more attractive option in a variety of new arenas.

Bibliography

1. Pazynyuk, T., Oreku, G.S., Li, J.: Distributed signature scheme establishment based on RSA. USENIX'08, USA, Poster Session (2008)
2. Pazynyuk, T., Li, J., Oreku, G.S.: Pros and cons in sensor networking. J. Acad. Rev., Russia **5**, 51–55 (2006)
3. Pazynyuk, T., Li, J., Oreku, G.S., Pan, L.: QoS as means of providing WSNs security. Seventh International Conference on Networking (ICN 2008), Mexico, pp. 66–71 (2008)
4. Pazynyuk, T., Li, J., Oreku, G.S.: Improved Feistel-based ciphers for wireless sensor network security. J. Zhejiang Univ. Sci. A, China **9**(8), 1111–1117 (2008)
5. Pazynyuk, T., Oreku, G.S., Li, J.: Mathematical model for wireless sensor nodes security. International Conference of Machine Learning and Cybernetics (ICMLC 2008), China, vol. 3, pp. 1305–1310 (2008)
6. Pazynyuk, T., Oreku, G.S., Li, J.: Distributed signature scheme (DSS) based on RSA. Inf. Technol. J., Pakistan **7**(4), 802–807 (2008)
7. Pazynyuk, T., Oreku, G.S., Li, J.: Reliable data aggregation protocol for wireless sensor networks. Third International Conference on Digital Information Management (ICDIM 2008), London, pp. 13–16 (2008)
8. Oreku, G., Li, J., Pazynyuk, T.: An application-driven perspective on wireless devices security. The Case of Distributed Denial-of-Service (DDoS) ACM/IEEE Proceeding WAiM Chania, Crete Island, Greece, pp. 81–83 (2007)
9. Oreku, G.S., Li, J., Pazynyuk, T., Mtenzi, F.J.: Modified s-box to archive accelerated GOST. J. Comput. Sci. Netw. Secur., South Korea **7**(6), 88–98 (2007)

© Springer International Publishing Switzerland 2016

G.S. Oreku and T. Pazynyuk, *Security in Wireless Sensor Networks*,
Risk Engineering, DOI 10.1007/978-3-319-21269-2

Zeitfracht Medien GmbH
Ferdinand-Jühlke-Straße 7
99095 Erfurt, Deutschland
produktsicherheit@kolibri360.de